TURBOCHARGERS

by Hugh MacInnes
Mechanical Engineer; Member, Society of Automotive Engineers,
Society of American Military Engineers,
American Society of Mechanical Engineers

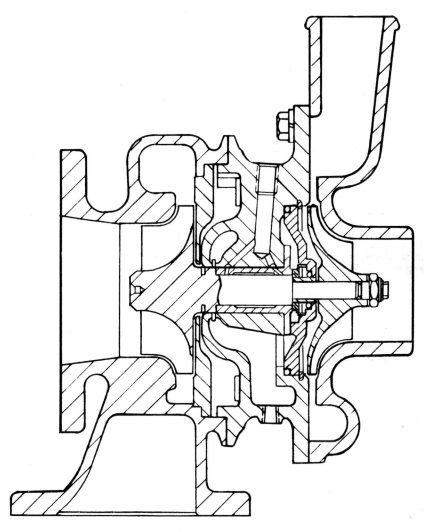

HPBOOKS
Published by the Penguin Group
Penguin Group (USA) Inc.
375 Hudson Street, New York, New York 10014, USA

Copyright © 1984 by Price Stern Sloan, Inc.
Cover photo: Dual turbocharged small-block Chevy by Gale Banks Engineering
Interior photos by Hugh MacInnes unless otherwise noted

Library of Congress Catalog Number: 76-6002
ISBN: 978-0-89586-135-1

PRINTED IN THE UNITED STATES OF AMERICA

55   54   53   52   51   50   49

NOTICE: The information in this book is true and complete to the best of our knowledge. All recommendations on parts and procedures are made without any guarantees on the part of the author or the publisher. Tampering with, altering, modifying or removing any emissions-control device is a violation of federal law. Author and publisher disclaim all liability incurred in connection with the use of this information.

CONTENTS

Introduction

Author Hugh MacInnes with one of his pet projects; a dual-turbocharged Oldsmobile.

In the introduction to my first book, *How To Select and Install Turbochargers,* I mentioned it started out as a 10-page pamphlet that kept growing. By the time it was published it was a 144-page book.

After three years, it was due for a revision. It soon became apparent that enough new items were being added to make it far more than a simple revision. So I started at the beginning and rewrote each chapter. Then I added five more.

More than five years passed since the first major revision. Two important things have occurred in that time.

First, the average passenger-car engine is now significantly smaller. Second, diesel engines have become much more important in the automotive scene. I've made a special effort to cover both of these developments in this revision.

Turbocharging engines with kits or custom installations has become popular worldwide. Although the sizes and shapes of cars vary throughout the world and the octane rating of available gasoline varies, the enthusiasm of the custom installers and kit manufacturers seems universal.

In my first book, I pointed out that only a few men had successfully turbocharged high-performance engines. This was in spite of the fact that their success should have sparked interest among many more.

Since that time, turbochargers have dominated tractor pulling, powerboat racing and auto racing on both circle and road-race tracks,

Turbos have "lost out" only in drag racing—probably because most people go to drag races for the noise. Because turbochargers tend to reduce noise, they may never become popular for drag racing.

Compared to adding power by the

Importance of turbocharging to the auto industry is exemplified by this '83 Indy 500 Buick pace car 4.1 liter-V6 engine. Maximum use of computer-controlled engine systems assisted engine in producing 400 HP in mild state of tune! Photo courtesy of Buick.

old methods of boring, stroking, special cylinder heads, cams, headers, and so on, turbocharging is the least-expensive way to get more usable horsepower from an engine.

One of the deterrents to turbocharging a stock engine in the past was the high compression ratio of the engine, particularly with high-performance V-8s. Very little turbocharging could be done without using an anti-detonant, even with the highest octane gasoline available.

As we all know, the advent of emission controls on passenger cars has caused a considerable drop in compression ratio to reduce combustion temperatures. This lowers the oxides of nitrogen in the exhaust and also allows using low-octane gasoline.

The low compression ratio is fine for turbocharging. With the availability of high-octane lead-free gasoline, it is possible to turbocharge these engines to at least 10 pounds per square inch (10 psi) boost pressure without any major modifications.

I hope the information contained in this book will accomplish three things:

First, enable the average automotive enthusiast to turbocharge his own engine with a reasonable chance of success.

Second, allow aftermarket kit manufacturers to avoid all the trial-and-error work necessary several years ago, when there was little previous knowledge on which to lean.

Third, be helpful to engine manufacturers who may consider turbocharging a small engine to do the job of a large one—without sacrificing fuel economy.

The tables of suggested turbocharg-

ers for various engines have been adjusted to include engines introduced since the last revision.

The hardest thing about writing a book is sitting down and doing the work of putting it together. On the other hand, it has given me the opportunity to become acquainted with many interesting people whom never would have met otherwise.

I've been associated with turbo chargers since 1952 and I can truthfully say there's never been a dull day. doubt if there are many other manufactured items that cover such a broad spectrum of mechanical engineering including fluid and thermodynamics metallurgy, lubrication, machine design, stress analysis, manufacturing techniques and internal-combustion engines.

1 Supercharging & Turbocharging

Top Fuel dragster takes advantage of instant boost from Roots blower and high efficiency of turbocharging. Photo by Tom Monroe.

Many discussions about engines and turbocharging assume auto enthusiasts are already familiar with the principles involved. As a result, many things that should be explained are left unsaid.

For this reason I am starting this chapter with the basic principles of operation of a typical four-stroke-cycle, spark-ignition engine. I will then explain how supercharging and, in particular *turbocharging*, affects its operation and output.

Standard automobile engines are naturally aspirated four-stroke-cycle, spark-ignition with four or more cylinders. The schematic of one cylinder, Figure 1-1, is familiar to all persons who have worked with auto-

mobile engines. It has the following sequence of strokes:

A. Intake—Fuel/air mixture is drawn through open intake valve.
B. Compression—Both valves are closed, charge compressed.
C. Power—Charge ignited by spark plug expands and pushes piston down.
D. Exhaust—Burnt gases expelled through open exhaust valve.

In addition to the number of cylinders, an engine is classified by its cubic-inch displacement, abbreviated CID. This is the number of cubic inches of air that will theoretically flow through a four-stroke engine during two complete revolutions of the crankshaft.

Engine displacement is also often described in cubic centimeters (cc) or liters (l). One liter is just over 61 cubic inches, but where both are listed on a chart in this book I have used 60 cubic inches to the liter to make the chart easier to read.

Volumetric Efficiency—In practice, the engine does not flow an amount of air equal to its displacement because:
1. There is always a slight pressure drop through the carburetor.
2. Intake ports and valves offer some restriction.
3. The exhaust stroke does not expel all burnt gases because some exhaust is trapped in the clearance volume.
4. The exhaust valves and exhaust pipes offer some restriction.

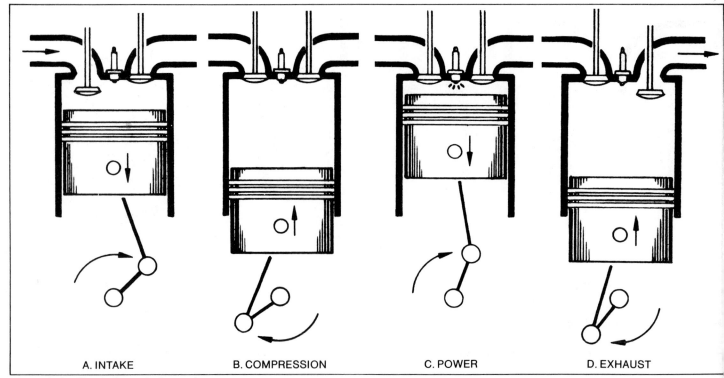

Figure 1-1—Simple four-stroke engine

| A. INTAKE | B. COMPRESSION | C. POWER | D. EXHAUST |

For these reasons, a normal automobile engine flows only about 80% of the calculated amount of charge. This amount is called *80% volumetric efficiency* or $\eta_{vol} = 80\%$.

It is possible to tune an engine to get higher volumetric efficiency by using the correct length intake and exhaust pipes for a given engine speed. This, coupled with oversized valves and ports and carefully designed intake and exhaust passages, make it possible to have an engine with a volumetric efficiency exceeding 100% at a certain speed. This is frequently done with racing engines but it is not practical for street use where a broad speed range is required.

SUPERCHARGING

Figure 1-2 shows a compressor added to the basic engine. This may be done either before or after the carburetor. In either case, if compressor capacity is greater than that of the engine, it will force more air into the engine than it would consume naturally aspirated. The amount of additional air will be a function of the intake-manifold-charge density compared to the density of the surrounding atmosphere. *Density* as used in this book is the weight of air per unit of volume. It is usually expressed as a percent or ratio.

There are two basic types of

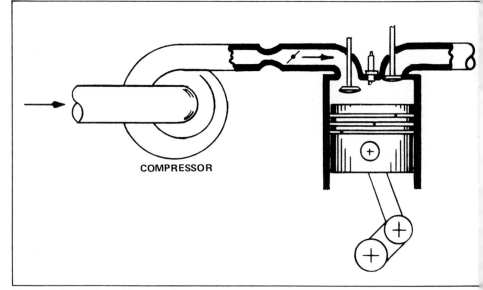

Figure 1-2—Engine with supercharger

compressors: positive-displacement and dynamic.

POSITIVE-DISPLACEMENT COMPRESSORS

Positive-displacement compressors, Figure 1-3, include reciprocating, lobe and vane compressors. There are also lesser-known types. Positive-displacement compressors are usually driven from the engine crankshaft through belts, gears or chains.

The compressor pumps essentially

the same amount of charge for each revolution of the engine regardless of speed. And, because it is a positive displacement device, all of this charge must pass through the engine. Assuming the compressor displacement is twice that of a normally aspirated engine, the intake-manifold pressure must rise to enable the engine to flow the same weight of charge delivered by the compressor. The manifold pressure will not be twice atmospheric.

This type of supercharger has the advantage of delivering approximately

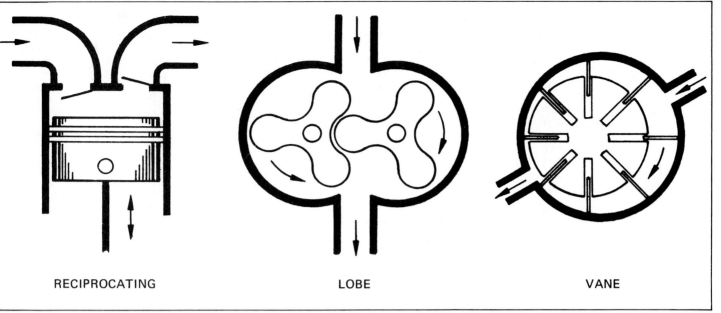

RECIPROCATING LOBE VANE

Figure 1-3—Positive-displacement compressors

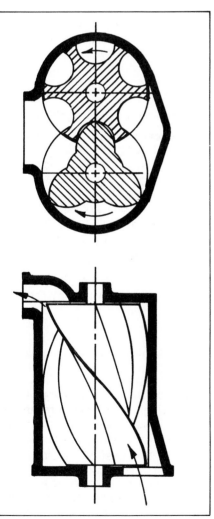

Figure 1-4—Lysholm compressor

the same manifold pressure at all engine speeds. The disadvantage is that crankshaft power is used to drive it.

The Roots-type lobe compressor also has the disadvantage of inherently low efficiency—below 50%. Low efficiency causes excessive charge heating and therefore higher thermal stress on the engine. The Lysholm-type lobe compressor, Figure 1-4, has much higher efficiency—up to 90%—but is extremely expensive and not practical for automotive use.

The reciprocating type has been used for many years on large stationary engines. Because it usually is attached directly to the crankshaft, it runs at crankshaft speed. It is rather large and cumbersome for use in an automobile engine.

The sliding-vane type is sealed internally by the vanes rubbing against the outer housing. Because of this, lubricating oil is usually mixed with the charge to prevent excessive wear on the sliding vanes. This lubricating oil lowers the fuel's octane rating.

An eccentric-vane type such as the smog pump used on many passenger-car engines does not require vane lubrication. Like the Lysholm type, this is very expensive in sizes large enough to supercharge most car engines.

Bendix Corporation made a version of the eccentric-vane compressor large enough for use on a passenger car. Bendix chose a design without built-in compression, Figure 1-6. This

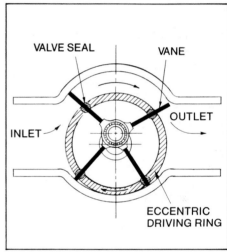

Figure 1-5—Eccentric-vane compressor

uses less power when not in operation. A cross section of this type compressor is shown in Figure 1-5 and a schematic of the system is in Figure 1-6.

DYNAMIC COMPRESSORS

All dynamic compressors are inherently high-speed devices. They obtain compression by accelerating the gas to a high velocity and then slowing it down by *diffusion*. Diffusion converts velocity energy into pressure energy by slowing down the gas without turbulence.

Dynamic compressors also come in several types.

An axial compressor, Figure 1-7, is basically a fan or propeller. With this type compressor it is difficult to

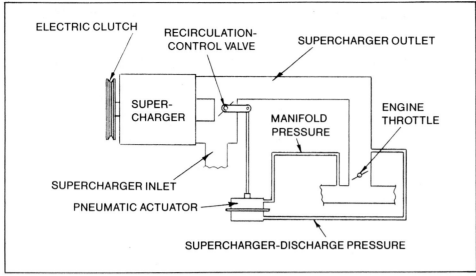

Figure 1-6—Bendix supercharger system

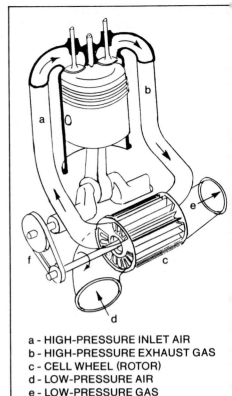

a - HIGH-PRESSURE INLET AIR
b - HIGH-PRESSURE EXHAUST GAS
c - CELL WHEEL (ROTOR)
d - LOW-PRESSURE AIR
e - LOW-PRESSURE GAS
f - BELT DRIVE

Figure 1-8—Comprex supercharger

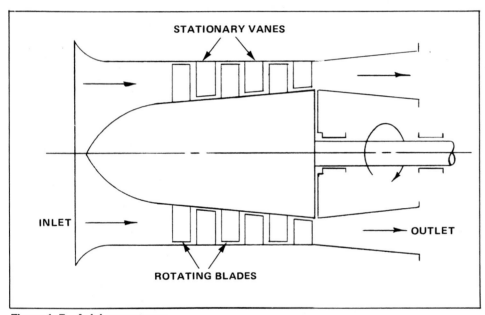

Figure 1-7—Axial compressor

obtain a pressure ratio much higher than 1.1:1 in a single stage. Higher ratios require several stages—one compressor feeding into another. The Latham axial supercharger fits this category.

Another type worth mentioning is the Comprex, Figure 1-8. This compressor uses the velocity of the exhaust gases to compress the charge air.

The cylinder is rotated at a speed that causes inlet air in the passage next to the intake manifold to be compressed by the exhaust gases. Air is then forced into the intake manifold. Because it uses the exhaust gases to compress the charge, supercharging is available immediately on demand.

The only *parasitic* loss of the Com-

prex is the friction in the drive system from the engine. Its drawback has been the high cost to eliminate leakage between the rotating and stationary elements.

A centrifugal-type compressor is shown in Figure 1-9 and discussed below.

Other types of dynamic compressors, such as mixed-flow and drag-type, are not ordinarily used to supercharge engines. I have not covered them in this book.

Centrifugal Compressors—The centrifugal-type compressor is shown in Figure 1-9. This differs from the axial-flow in that the gas direction is changed approximately 90°.

Also, the air is in contact with the

compressor impeller for a longer period of time per stage. The longer the contact with the impeller, the greater the acceleration. Consequently, it is possible to achieve a considerably higher pressure ratio in a single stage of a centrifugal compressor. A 4:1 pressure ratio is not uncommon.

Because the centrifugal compressor must be driven at very high speed, it is difficult to drive from the crankshaft. As can be seen from the *compressor map*, Figure 1-10, a compressor capable of supplying a 3:1 pressure ratio with a capacity large enough for a 350-CID engine must run at nearly 115,000 rpm. It would require a step-up gear with a 23:1 ratio on an engine running at 5000 rpm. Read about compressor maps, page 10.

This is impractical for a number of reasons. The cost of the transmission is one. Also, sudden changes in engine speed during shifting require a slip clutch or other cushioning device. Otherwise the speed change would destroy the supercharger-drive gears.

During the 1920s, gear-driven centrifugal superchargers were used with success on race cars turning out as high as 300 HP from 90 CID engines.

That was 3.33 HP per cubic inch. Because the gear-driven supercharger probably used about 60 HP from the crankshaft, the same engine equipped with a turbocharger would have produced about 360 HP—4 HP per cubic inch.

As of 1981, turbocharged race engines were producing about 900 HP from 160 cubic inches, about 5.6 HP per cubic inch. That's a 40% improvement in 45 years.

Engine builders back in those days lacked the high-strength materials we have today. So, they cooled the charge between the compressor and the engine. This reduced both the thermal and mechanical loads while increasing the output. Advantages of intercooling are discussed in Chapter 11.

The biggest disadvantage of the centrifugal compressor when used as a supercharger is: Pressure output varies considerably with compressor speed.

Looking again at the typical map in Figure 1-10, we see this particular compressor puts out 1.2:1 pressure ratio at 40,000 rpm. This represents approximately 3 pounds boost pressure at sea level. This same compressor will produce 1.92:1 pressure ratio at 80,000 rpm, 13.8 pounds of boost pressure. In this particular case, doubling the speed increased the boost pressure over four times.

A rule of thumb is: Boost pressure increases as the square of the speed of the compressor.

About the only way to overcome this problem with an engine-driven centrifugal compressor is to have a variable-speed drive. The Paxton supercharger that came out in the early '50s had a variable V-belt drive actuated by the accelerator pedal. When the accelerator was in normal driving position, the supercharger ran at a relatively low speed. When the driver pressed the accelerator to the floor, an actuator decreased the supercharger pulley's effective size. It caused the supercharger to run much faster. This system worked well but was an added complication.

In addition to its higher overall efficiency—some centrifugal compressors operate at better than 80% efficiency—the centrifugal compressor has another advantage over the positive-displacement compressor. Because it is not a positive device, it can withstand a small backfire

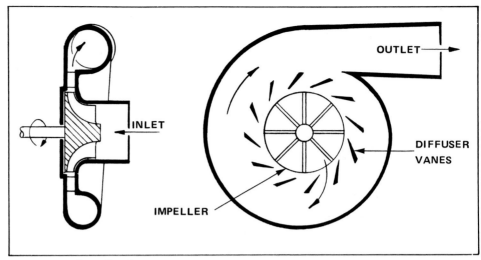

Figure 1-9—Centrifugal compressor

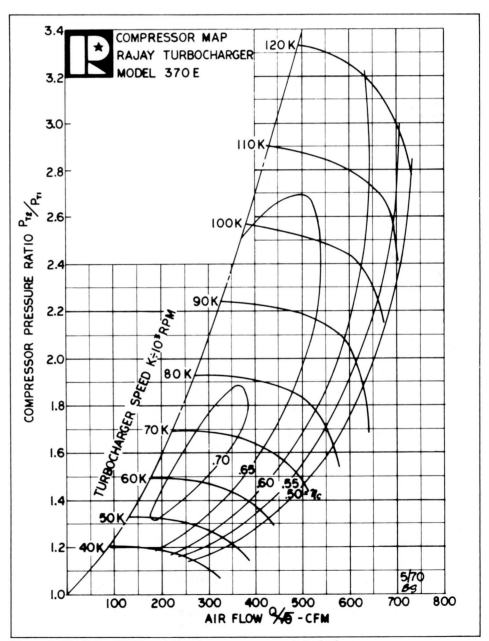

Figure 1-10—Rajay 300E centrifugal-compressor map is typical

CENTRIFUGAL-COMPRESSOR MAP

Centrifugal compressors differ in many respects from positive-displacement types. The operating characteristics of a particular compressor must be known so the user can match it to the engine.

The operating characteristics are shown in a graph or *compressor map*, Figure 1-10. The efficiency curves are like "hills" on a topological map. Efficiency increases from right to left until it reaches a peak-efficiency "island." Peak efficiency is usually somewhere around 70%, or 0.70 as shown in Figure 1-10.

The typical map is the result of running the compressor on a test rig, Figure 1-11. The compressor test stand is instrumented and the following data is recorded:

1. Shaft speed
2. Compressor inlet pressure
3. Compressor outlet pressure
4. Compressor inlet temperature
5. Compressor outlet temperature
6. Compressor flow.

Holding the compressor speed constant, air flow is reduced by throttling the compressor discharge. Where the compressor outlet pressure and flow become unsteady is called the *surge point.* This surge point determines the minimum usable flow at that shaft speed. Below this air flow the compressor will not operate without pulsing or *surging.*

The flow is then increased until the compressor-discharge gage pressure is about one-half the gage pressure at the surge point. This is usually referred to as the *choke condition* and is above the maximum usable flow of the compressor.

Several intermediate flow points are run so a smooth curve can be plotted through them.

There are two ways to measure pressure—*static* and *total,* Figure 1-12. Static pressure is not affected by the velocity of the gas. Total pressure consists of static pressure *plus* the velocity pressure of the gas. Consequently, total pressure is always greater than static pressure.

The actual curve plotted on the map is the calculated pressure ratio, P_{T2}/P_{T1}, based on total pressure. Total pressure at the compressor inlet (P_{T1}) and outlet (P_{T2}) determines the pressure ratio.

Compressors are usually rated by pressure ratio, not by inches of mercury or pounds per square inch.

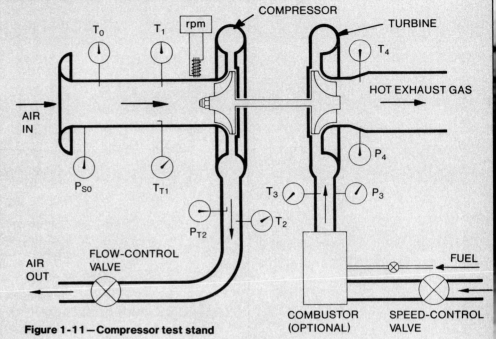

Figure 1-11—Compressor test stand

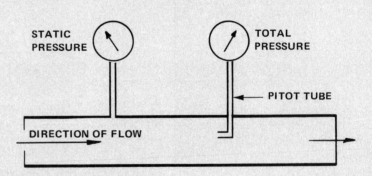

Figure 1-12—Static and total pressure

Pressure ratio, P_{T2}/P_{T1}, is independent of inlet pressure. It remains the same regardless of barometric pressure, staging or other factors that can affect actual boost pressure.

The speed lines are run at various speeds, usually in increments of 10,000 rpm. Some aerodynamicists insist on running compressors at particular rotor-tip speeds, which results in rotational speeds such as 96,250 rpm.

These maps usually show air flow in pounds per minute. But the reciprocating engine is a displacement machine, not a mass-flow machine. Consequently, pounds per minute must be converted to cubic feet or cubic meters per minute.

On most maps there will be the note, rpm $\div \sqrt{\theta}$ next to the surge line. This is a speed-correction factor for compressor-inlet air temperature. The factor is rarely used except by the people who make the maps and in cases where the temperature is very low (−65F), as on an airplane operating at high altitude.

After the runs are completed and the pressure ratios and efficiencies calculated (Chapter 2), the results are plotted, Figure 1-10. Although this particular map has efficiency lines drawn down to 50%, most turbochargers are matched to operate between the surge line and 60% efficiency.

Chapter 4 describes how to use these maps when matching a turbocharger to an engine.

through the intake system without damage. A backfire on a turbocharged engine is no worse than a naturally aspirated engine.

This is not so with a positive-displacement compressor. A *small* backfire can usually be handled by pop-off safety valves mounted somewhere between the supercharger and the engine. A *large* backfire may remove the supercharger completely from the engine.

Because of the inherent high speed of the centrifugal-type compressor, the size and weight of the unit are considerably less than the positive-displacement type. A complete turbocharger system capable of enabling an engine to produce over 1000 HP weighs only about 25 pounds.

TURBOCHARGERS

Driving a centrifugal compressor would always be a problem except that a turbine is also a high-speed device. For this reason, we can couple them directly together without the use of gears. The turbine is driven by the exhaust gases of the engine. This utilizes energy usually dumped overboard in the form of heat and noise. Exhaust gases are directed to the turbine wheel through vanes or a nozzle in turbine housing, Figure 1-13.

Many people think this exhaust-gas energy is not free because the turbine wheel causes back pressure on the engine exhaust system. This is true to a certain extent, but when the exhaust valve first opens, the flow through it is critical. Critical flow occurs when the cylinder pressure is more than twice the exhaust-manifold pressure. As long as this condition exists, back pressure will not affect flow.

After cylinder pressure drops below critical pressure, exhaust-manifold pressure will definitely affect the flow. Higher cylinder pressure of the turbocharged engine during the latter portion of the exhaust stroke will still require some crankshaft power.

When an engine is running at wide-open throttle with a well-matched high-efficiency turbocharger, intake-manifold pressure will be considerably higher than exhaust-manifold pressure. This intake-manifold pressure will drive the piston down during the intake stroke, reversing the process of the engine driving the gases out during the exhaust stroke.

During the overlap period when both valves are open, the higher

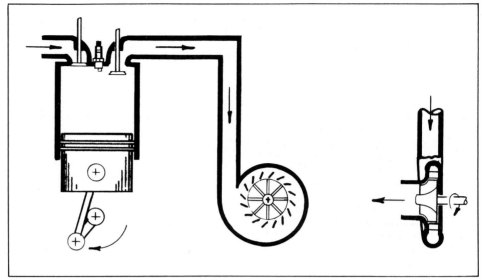

Figure 1-13—Engine-driven gas turbine

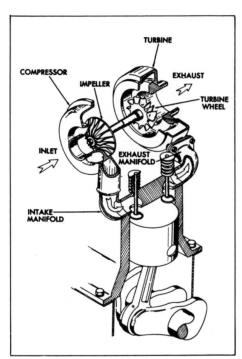

Air and exhaust flow through turbocharger. Drawing courtesy of Schwitzer.

intake-manifold pressure forces residual gases out of the clearance volume, scavenging the cylinder. Intake-manifold pressure as much as 10 psi higher than exhaust-manifold pressures have been measured on engines running at about 900 HP. Good scavenging can account for as much as 15% more power than calculated from the increase in manifold pressure of the naturally aspirated engine.

Exhaust-gas temperature will drop as much as 300F (133C) when passing through the turbine. This temperature drop represents fuel energy returned to the engine by the turbocharger.

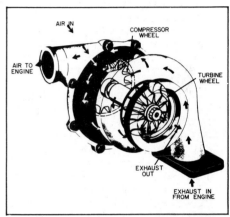

Exhaust enters turbine housing tangentially and exits axially; air is drawn into compressor housing axially and exits tangentially. Drawing courtesy of Schwitzer.

Under ideal conditions, intake-manifold pressure at full throttle in a naturally aspirated engine is the same as ambient—outside—air pressure. At sea level, ambient air pressure is considered to be 14.7 pounds per square inch (psi), or 29.92 inches of mercury (in.Hg).

Notice I said "ideal." Any restriction in the induction system reduces the amount of pressure available to each cylinder.

For example, a 1-psi drop through the carburetor means that pressure in the intake manifold is 1 psi less than ambient. At sea level, this means only 13.7 psi of pressure is available to charge the cylinders.

SUMMARY

For a given type of fuel, more power can be obtained from an engine by turbocharging than by any other method.

2 Turbocharger Design

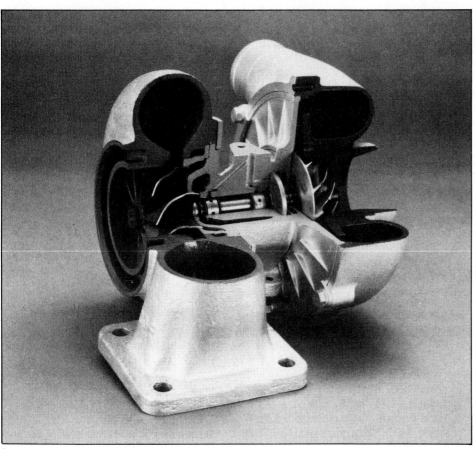

Internal configuration of Warner-Ishi RHB7

The basic function of the turbocharger is essentially the same as the first one designed by Alfred Büch many years ago. The mechanical design is simpler and size for a given output is much smaller. And in spite of the trend toward higher prices for everything, the price of a turbocharger per horsepower increase is much less now than it was in the 1950s.

Until 1952, most turbochargers used ball or roller bearings and an independent oil system with built-in pump. In addition, they were water-cooled. Today's units use floating-sleeve bearings lubricated by the engine's oil and pump. They are cooled by a combination of oil and air, and engine coolant, in some cases.

Turbocharger design varies from one manufacturer to another. Basically, all production models have a compressor on one end and a turbine on the other, supported by bearings in between, Figure 2-1.

There are seals between the bearings and the compressor and also between the bearings and the turbine. These prevent high-pressure gases from leaking into the oil-drainage area of the bearing housing and eventually into the crankcase of the engine. Seals are much better known for keeping oil from leaking into the compressor or turbine housing. How well they do this job often depends on the installation.

COMPRESSOR DESIGN

The centrifugal compressor consists of three elements that must be matched for optimum efficiency: impeller, diffuser and housing.

The impeller rotates at very high speeds. Gas passing through it is accelerated to a high velocity by centrifugal force.

The diffuser acts as a nozzle in reverse to slow down the gas without turbulence. Slowing down the gas causes it to increase in pressure and, unfortunately, in temperature.

The housing around the diffuser is used to collect this high-pressure gas and direct it to wherever it is used. In some cases, the housing itself is also a diffuser.

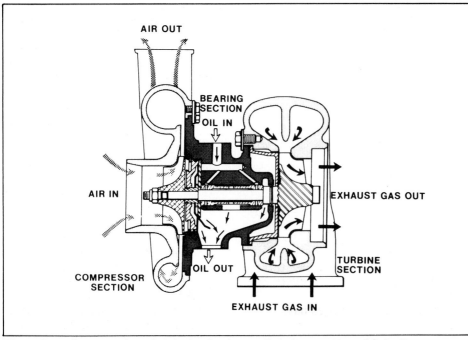

Figure 2-1 — Cross section of typical turbocharger. Drawing courtesy of Schwitzer.

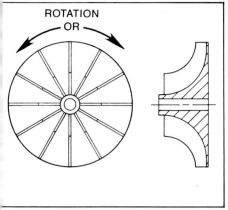

Figure 2-2—Simple compressor impeller

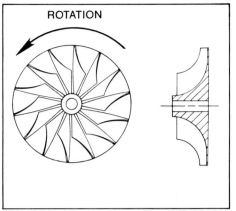

Figure 2-3—Impeller with curved inducer

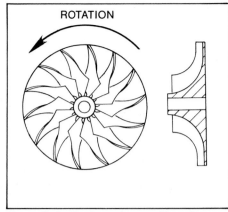

Figure 2-4—Backward-curved impeller

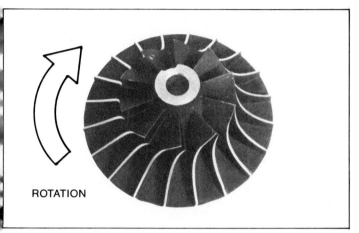

Impeller with curved inducer

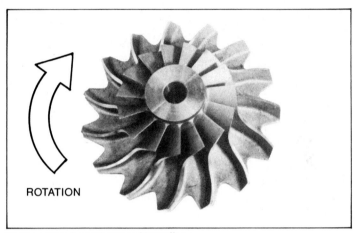

Photo of backward-curved impeller

Impellers—Over the years, the design of compressor impellers used in superchargers has varied considerably due to "state of the art" in the thermodynamic design of compressors and in manufacturing techniques.

Figure 2-2 shows a simple 90° straight-blade impeller with no curved inducer section. All blade elements lie on straight lines that pass through the center of the impeller hub. This is often referred to as a *90° radial wheel*. It has not become popular because of its relatively low efficiency caused by shock losses at the inlet. This shape is relatively easy to produce by die casting, permanent-mold casting, plaster casting, or even milling.

Figure 2-3 shows a similar impeller, but with curved inducer blades. In this design, the blade elements are not radial, but are curved away from the direction of rotation. The angle of curvature at the inlet of the inducer blades is designed so the air entering the impeller will be at approximately the same angle as the blades, thereby reducing inlet losses to a minimum.

Originally, this type of wheel was expensive to cast because it required a separate plaster core for each gas passage. These cores were pasted together by hand to make the final mold. In recent years this type of compressor impeller has been cast by the *investment* or *lost-wax* method.

When an impeller or "wheel" is cast by this method, a die is made similar to that for die casting except that wax is cast into the die rather than metal. The wax is removed from the die and covered with liquid plaster.

After the plaster hardens, it is heated to remove the wax by melting. Molten aluminum alloy is then poured into the cavity. The plaster is then broken away to expose the newly cast impeller. This process makes smooth, high-strength impellers but is still expensive.

More recently, foundries have been using the *rubber-pattern process*.

In this method, a die similar to the wax die is constructed. But, instead of being filled with molten wax, it is filled with a rubber compound that solidifies in the die. After it is removed from the die, this rubber pattern is covered with liquid plaster. It is allowed to harden the same as with the wax pattern. At this point, the process differs in that the flexible rubber pattern can be removed from the plaster after it hardens. After the rubber pattern is removed from the plaster, it returns to its original shape and may be used again.

The molten aluminum alloy is poured into the cavity. Then the plaster is broken away to release the newly cast impeller. This method of casting has made possible the use of compressor-impeller shapes that were not considered economical from a casting viewpoint a few years ago.

Figure 2-4 shows what is known as a *backward-curved compressor impeller*. In this design, the blade elements are not radial but actually curve backward from the direction of rotation.

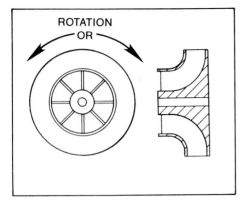

Figure 2-5—Shrouded impeller

Wheels of this type produce very high efficiency but do not have as high a pressure ratio for a given diameter and speed as the 90° radial wheels. Strength is inherently less than that of the 90° radial wheel because the centrifugal force at high speed tends to bend the blades at their roots. Because of the lower pressure ratio for a given speed and the inherently lower strength of this type wheel, it is not normally used at pressure ratios above 3:1.

The use of turbochargers with only 7-psi boost on passenger cars has renewed the popularity of the backward-curved impeller.

Figure 2-5 shows a shrouded impeller. This design is certainly the most expensive to manufacture and the weakest of all the designs because the blades must carry the weight of the shroud as well as their own.

Maximum efficiency of a shrouded impeller is usually very high because there is minimal recirculation from the impeller discharge back to the inducer. The low strength, high cost and tendency for the shroud to collect dirt have eliminated the use of the shrouded impeller in automotive use.

In 1952, the turbocharged Cummins diesel-powered race car, which ran in the Indianapolis 500, had to retire from the race due to dirt buildup on a shrouded impeller.

In the late 1950s when shrouded impellers were used on construction equipment, the service manual included a preventive-maintenance procedure showing how to flush soapy water through the compressor to remove dirt buildup on the shroud.

Different contours can be machined on the cast wheels. This allows using both turbine-wheel and compressor-impeller castings for more than one flow-size turbocharger.

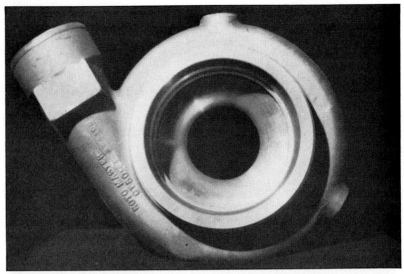

Figure 2-6A—Scroll diffuser

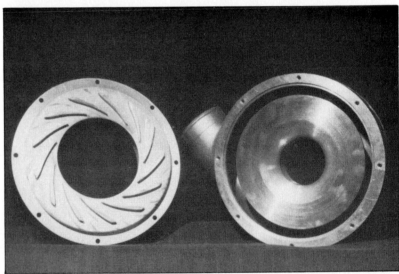

Figure 2-6B—Vaned diffuser

Figure 2-6C—Parallel-wall diffuser

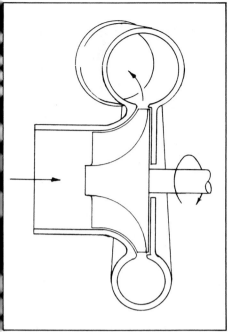

Figure 2-7—Scroll diffuser

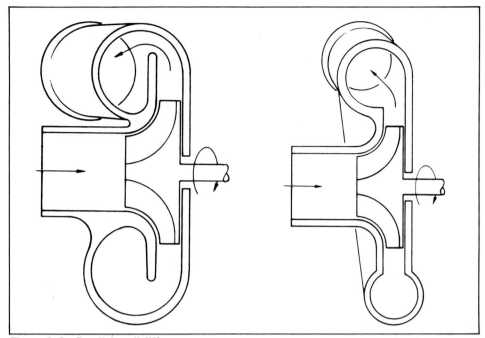

Figure 2-8—Parallel-wall diffuser

The turbine is not as sensitive to flow changes as the compressor, so it is common for a given turbocharger model to have many more compressor-impeller variations than turbine-wheel variations. Compressor maps usually specify both the blade-tip height and the inducer diameter of the impeller. The compressor can then be identified even if the part number has been obliterated.

Diffusers—The diffuser increases the *static* pressure of the gas in the compressor.

The difference between static and total pressures is shown schematically in Chapter 1, Figure 1-12. A static-pressure probe is not affected by the velocity of the gas. A total-pressure probe measures static pressure *plus* the velocity pressure of the gas. Consequently, total pressure is always greater than static pressure.

Three types of diffusers are normally used with centrifugal compressors, Figure 2-6. They may be used singly or in combination with each other. The simplest is the scroll type, Figure 2-7. It consists of a volute or snail shape around the outside of the impeller.

The cross-sectional area of the scroll increases in proportion to the amount of air coming from the impeller. When designed correctly, it slows the gas down and converts velocity energy into pressure energy.

Figure 2-8 shows a parallel-wall

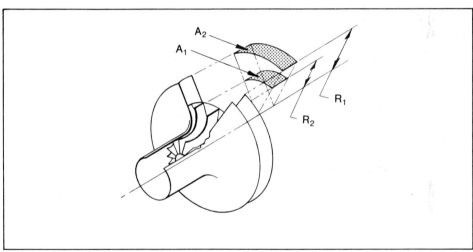

Figure 2-9—Area increase of parallel-wall diffuser. Drawing by Tom Monroe.

diffuser. It has an increase in area from the inside diameter of the diffuser to the outside diameter proportional to these two diameters, Figure 2-9. In other words, if R_2 is twice as great as R_1, then A_2 is twice as great as A_1. Assuming the gas were flowing in a radial direction, the velocity at R_2 would be half that at R_1. The gas actually flows in a spiral rather than a purely radial direction. But regardless of this, the gas velocity at the outer diameter of the diffuser is considerably less than at the inner diameter.

Figure 2-10 is a compressor with a vane-type diffuser. The vanes are designed so their leading edges will be in line with the direction of gas flow from the impeller. From this point, vane curvature will force the gas to flow and be slowed down in a way that favors a particular engine speed or torque requirement. Vanes may be used to alter the pressure ratio and flow characteristics of the compressor.

Compressors with vane-type diffusers normally have a high peak efficiency. They frequently have a narrower *range* than compressors with a vaneless diffuser.

Range refers to the range of engine displacements on which a given compressor may be used. A compressor with a broad range can be adapted to a wider engine-displacement spread than one with a narrow range. When installed on a specific engine, a turbocharger with a wide range is effec-

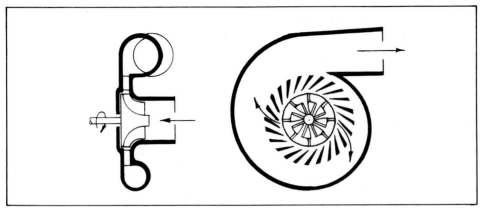

Figure 2-10—Vaned diffuser

tive over a broader rpm band.

The surfaces of the compressor impeller, the diffuser and the compressor housing are made as smooth as is economically practical. Any roughness may cause some of the gas to detach itself from the surface, causing eddy currents which reduce the overall efficiency of the compressor. The accompanying photos show examples of various types of compressor housings and diffuser.

COMPUTING INTAKE-CHARGE TEMPERATURE

As mentioned earlier, a compressor will always raise the temperature of the gas as it raises pressure. With a centrifugal compressor operating at 100% efficiency, the formula for this temperature increase is:

$$T_2 = T_1 \left(\frac{P_2}{P_1}\right)^{0.283}$$

Where
T_1 = Inlet temperature °R
T_2 = Outlet temperature °R
°R = °F + 460
P_1 = Inlet pressure absolute (ABS)
P_2 = Outlet pressure ABS.

Inlet pressure is usually barometric, which means outlet pressure is barometric *plus* gage pressure. The exception to this is in *staged* applications.

In a staged application the inlet pressure of the *first* turbocharger is barometric. The inlet pressure of the second turbocharger is the outlet pressure of the first turbocharger. This is explained further in Chapter 16, Tractor Pulling.

Example: Assume inlet temperature of 70F.
Then
$$T_1 = 70 + 460$$
$$= 530R$$

Assume inlet pressure is 0 psig.

Then
$$P_1 = 0 + \text{barometer}$$
$$= 0 + 14.7 \text{ psia}$$
$$= 14.7 \text{ psia (lb per sq in. abs)}$$

Assume outlet pressure is 17 psig. Then
$$P_2 = 17 + \text{barometer}$$
$$= 17 + 14.7 \text{ psia}$$
$$= 31.7 \text{ psia}$$

The theoretical outlet temperature T_2 is:
$$T_2 = (70 + 460) \times \left(\frac{17 + 14.7}{14.7}\right)^{0.283}$$
$$T_2 = 530 \, (2.16)^{0.283}$$
$$T_2 = 530 \times 1.243$$
$$T_2 = 659R \text{ (or 199F — a temperature rise of 129F).}$$

Compressor Efficiency—The above calculation assumed 100% *adiabatic* efficiency, or done with no gain or loss of heat. That's about as obtainable as perpetual motion. Lower efficiency increases the temperature of the compressed air still further.

Compressors referred to in this book are capable of putting out about 70% efficiency. Although this is very commendable for compressor impellers of 3.0-inch diameter, it tends to increase the temperature of the compressed air still further. At 70% adiabatic efficiency the actual temperature rise is computed:

$$\frac{\text{Ideal Temp. Rise}}{\text{Adiabatic Eff.}} = \text{Actual Temp. Rise}$$

In this case,
$$\frac{127F}{0.7} = 181F.$$

Adding this to the compressor inlet temperature,

$$70F + 181F = 251F$$

In terms useful to the end user, a 70%-efficient supercharger compressor producing 17-psi boost at sea level on a 70F day will produce an intake-manifold temperature of 251F.

This series of calculations may look like a lot of work but most of it can be eliminated by the use of Table 1.

With the table you can calculate discharge temperature directly with simple addition and multiplication. Assume the following conditions:

Inlet Temperature = 80F
Pressure Ratio r = 1.9
Compressor Efficiency η_c = 0.65

From Table 1 where
$$r = 1.9$$
$$Y = 0.199$$

Ideal Temperature Rise
$$\Delta T_{ideal} = T \times Y$$
$$= (460 + 80) \times 0.199$$
$$= 107.5F$$

Actual Temperature Rise
$$\Delta T_{actual} = \frac{\Delta T_{ideal}}{\eta_c}$$
$$= \frac{107.5}{0.65}$$
$$= 165.4F$$

Compressor-Discharge Temperature
$$T_2 = T_1 + \Delta T_{actual}$$
$$= 80 + 165.4$$
$$= 245.4F$$

TABLE 1							
r	Y	r	Y	r	Y	r	Y
1.1	0.27	1.6	0.142	2.1	0.234	2.6	0.311
1.2	0.053	1.7	0.162	2.2	0.250	2.7	0.325
1.3	0.77	1.8	0.181	2.3	0.266	2.8	0.338
1.4	0.100	1.9	0.199	2.4	0.281	2.9	0.352
1.5	0.121	2.0	0.217	2.5	0.296	3.0	0.365

r = pressure ratio
$Y = r^{0.238} - 1$

This means if you start with air at 80F, compress it to a pressure ratio of 1.9 with 65% efficiency, you will end up with air at 245.4F. If this sounds bad, a Roots-type blower with 45% efficiency will produce a temperature of 319F.

The Y Table in Table 1 is for demonstration purposes only. A more detailed table in the appendix has pressure ratios up to 10:1. A few years ago, 3:1 pressure ratio was considered more than adequate. The tractor-pull enthusiasts now run manifold pressures over 200 psig in their contests. That is not a misprint. *It is actually over 200 psig intake-manifold pressure!*

If this method of calculating intake-manifold temperature is still too complicated, it can be done without any calculations by using Don Hubbard's chart at right.

TURBOCHARGER RANGE

When discussing centrifugal compressors, the terms *broad range* or *narrow range* are often used. As mentioned earlier, a compressor with a broad range can be adapted to a wide range of engine displacements.

These terms could have several meanings, but in turbocharger work, *range* is normally the width of the compressor map at about 2:1 pressure ratio. The width of a compressor map is taken from the surge line to the 60% efficiency line. This is the area where the compressor will be used. Figure 2-11 includes examples of compressors with narrow, normal and broad ranges.

The surge area on the left side of each map is a region of pressure and flow where the compressor is unstable. Depending on the compressor, this instability will vary from a sharp banging sound, to a slush-pump-like action, to no surge at all. Normally, the narrower the range of the compressor, the sharper the surge.

Why not design all compressors with extremely broad ranges? This would be fine except that as a general rule, the broader the range, the lower the peak efficiency. Compressors designed to operate at only one flow have a very narrow range and a high peak efficiency. One example is a compressor used on an industrial gas turbine.

Turbocharger compressors, on the other hand, must not only work over

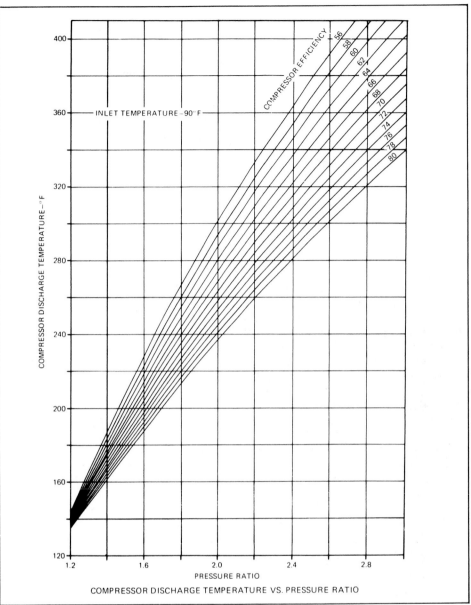

Compressor-discharge temperature vs. pressure ratio by Don Hubbard

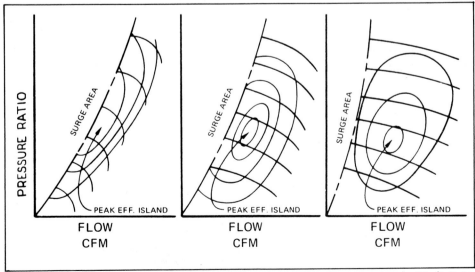

Figure 2-11—Compressor maps of three different centrifugal compressors with (left to right) narrow, normal or medium, and broad ranges

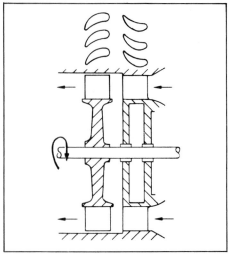

Figure 2-12—Axial-flow turbine and nozzle

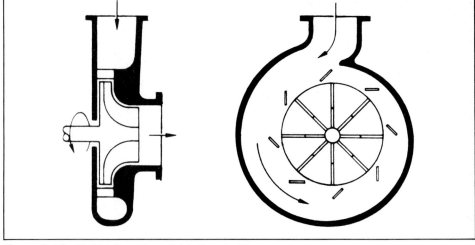

Figure 2-13—Radial-flow, or radial-inflow, turbine

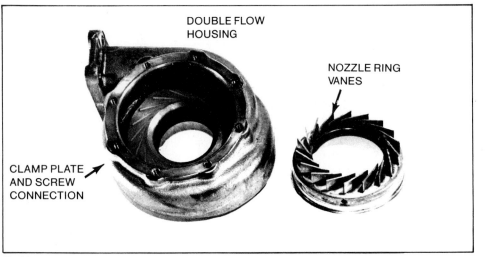

DOUBLE FLOW HOUSING

NOZZLE RING VANES

CLAMP PLATE AND SCREW CONNECTION

Turbine housing with multiple nozzle vanes on a removable ring. Photo courtesy of Schwitzer.

the broad operating range of a particular engine, but must be used on engines of many different sizes. This is so a different compressor doesn't have to be designed for each engine.

To compromise, a turbocharger compressor will normally have a *flow range* of about 2:1. This means maximum useful flow at 2:1 pressure ratio will be twice the flow at the surge limit. This compromise of peak efficiency and mild surge makes a particular turbocharger adaptable to a wide range of engine displacements.

TURBINE DESIGN

Axial-Flow Turbines—Textbooks on turbine design normally spend several chapters on turbines like that shown in Figure 2-12. This design, known as an *axial-flow turbine,* has been used for almost 100 years on large steam and gas turbines.

Axial-flow turbines use stationary nozzle vanes to direct the gases at an angle to make the turbine rotate. The distance between the stationary nozzle vanes and the turbine blades is critical.

The axial-flow turbine is used in most cases where the radial wheel is too large to cast in one piece. The normal method of manufacture uses a disk of high-strength material with the turbine blades attached either mechanically or by welding. Turbine blades are made of material with great heat and corrosion resistance.

Radial-Inflow Turbines—Radial-inflow turbines are either casually referred to in textbooks or ignored completely. But they are used almost exclusively in turbochargers with capacities up to about 1000 HP. The radial-inflow turbine shown schematically in Figure 2-13 is more economical to produce than an axial-flow turbine in small sizes.

The radial-inflow turbine also has a nozzle to direct the flow of gases to the turbine wheel at the best possible angle. The distance between the trailing edge of the nozzle vanes and the turbine wheel is not as critical as for the axial-flow turbine.

This is because gases continue at approximately the same angle as directed by the nozzle vanes. This spiral flow of fluid—a free vortex—can be observed any time you watch water go down a drain. Because of this phenomenon, the number of nozzle vanes is not critical on a radial-inflow turbine.

By designing the turbine housing in a scroll or volute shape, only one nozzle vane is needed. This considerably reduces the cost of building a radial-inflow turbine, although it is necessary to change turbine housings rather than nozzles when the turbine is used in different conditions. Changing turbine housings may be awkward but it is not expensive. A small turbine housing does not cost any more to manufacture than a small nozzle. This is not true with a large turbine. Turbochargers with 7-inch-diameter or larger turbines usually have a separate nozzle with many vanes such as that pictured above.

Turbine Sizing—The *nozzle area* of either an axial- or radial-inflow turbine with multiple nozzle vanes is the cross-section of a single nozzle opening multiplied by the number of vanes. In a turbocharger, the larger the nozzle area, the slower the turbocharger will run.

For example, if a turbocharger has

a 1.0-square-inch nozzle and is putting out too much boost at a given condition, the nozzle may be changed to one with 1.2-square-inch area. This will slow down the turbine and reduce the boost from the compressor. If this is not enough, the area may be increased to 1.4 square inches for a further reduction in speed and, therefore, boost.

Nozzle area is measured a little differently when the single-opening or vaneless turbine housing is used. In this case, the area alone shown as A in Figure 2-14 will not necessarily determine the amount of gas flow into the turbine. But A divided by R—the distance from the center of the turbine wheel to the centroid of area A—will determine gas flow for a given turbine wheel.

If A is increased, the turbine will slow down in the same manner as one with a multi-vane nozzle. If a housing is used with a larger R, as shown in Figure 2-15, A must be increased to retain the same A/R ratio.

This is important to the turbine designer, not to the user, because the variations in R from housing to housing are insignificant.

Not all turbocharger manufacturers use the A/R method of sizing turbine housings. Schwitzer uses area A only and because of this, their turbine housing sizes are not directly comparable to those of AiResearch, Rajay, Roto-Master or Warner-Ishi (IHI). This ratio or area is usually cast or stamped into the turbine housing by the manufacturer.

The thing that is important to the user is to know the housing A/R. If he has a turbocharger with a turbine housing A/R of 0.7 and he wants to run his turbocharger slower, he knows that a turbine housing designed to fit this same turbine wheel with an A/R of 0.9 will definitely cause his turbocharger to run slower. On the other hand, if he wants it to run faster, he knows a turbine housing with an A/R of 0.6 or 0.5 will cause the turbocharger to run faster and give more boost. The same is true with area alone in the Schwitzer vaneless turbine housing or any turbocharger with a multi-vaned nozzle.

This is discussed further in the chapter on sizing and matching and also in the chapter on kits. On most turbochargers, changing the turbine housing is a simple task involving a

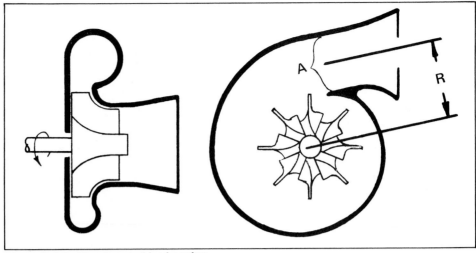

Figure 2-14—Vaneless turbine housing

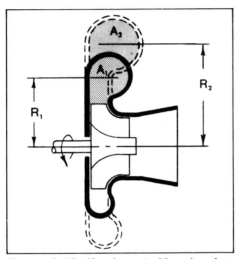

Figure 2-15—Vaneless turbine housing with larger R

Figure 2-16—180° divided turbine housing

few bolts or a V-band clamp.

Other Design Factors—Turbine designers try to use the pulse energy of gases coming from the individual cylinders to increase the boost pressure at very low speeds. On large axial-flow turbines this is done by running a separate exhaust stack from each cylinder to the turbine nozzle.

This was not practical on a radial-inflow turbine, so housings were divided in 180° increments, Figure 2-16. This made some improvement at very low engine speeds but the gas does have a tendency to reverse its flow due to centrifugal force when there is no high-pressure pulse in the housing.

Turbine-housing designs by John Cazier (patents 3,292,364 and 3,383,092), D. H. Connor (3,270,495), and Hugh MacInnes (4,027,994), divide the turbine housing axially, Figure 2-17. This prevents

the gas pulses from reversing and is used on engines where high torque is desired at low engine speed.

Different contours can be machined on the cast wheels. This allows using both turbine-wheel and compressor-impeller castings for more than one flow-size turbocharger. Figure 2-18 shows three different contours on both turbine and compressor castings.

Because the turbine is not as sensitive to flow changes as the compressor, it is common for a given turbocharger model to have many more compressor-impeller variations than turbine-wheel variations. Compressor maps usually specify both the blade-tip height and the inducer diameter of the impeller. The compressor can then be identified even if the part number has been obliterated.

Turbine-Housing Materials—Years ago, most turbine housings were made from Ni-Resist ductile cast iron.

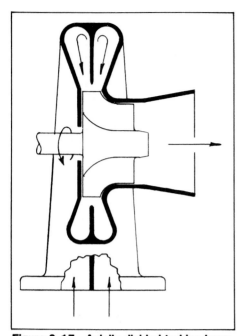

Figure 2-17—Axially divided turbine housing

Jim Kinsler used axially divided turbine housing to improve low-end response on this road-racing Corvette.

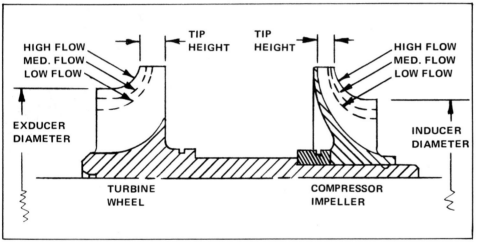

Figure 2-18—Contour relationship for different flows

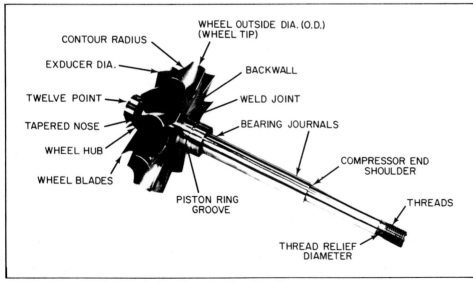

Construction details of typical turbine wheel and shaft

This cast-iron alloy contains between 20% and 30% nickel. It is still recommended for engines run continuously at extremely high temperature, as on a track-type race car or an airplane. It is also more resistant to corrosion from sea water than other types of cast iron, so it's also recommended for marine use.

The high cost of nickel has made it almost prohibitive to use Ni-Resist in diesel-engine turbochargers or those designed for intermittent use on passenger cars.

For many years turbine housings for diesel and passenger-car applications were made from a cast iron similar to that used in exhaust manifolds of automobiles and diesel engines. This material worked fairly well.

Around 1960, ductile cast iron became available at almost the same cost as gray iron. Ductile iron not only machines well, but can be welded more readily than gray iron. Most turbine housings are now ductile iron.

Turbine & Shaft Materials—Some of the earlier automotive-type turbine wheels were made by machining the blades from solid forgings. The curved or exducer portion—outlet section of the blades—was cast from a material such as Hastelloy B. The exducer was then welded to the turbine wheel and the assembly bolted to the shaft. This method was expensive and time-consuming, so better methods had to be evolved for mass production.

For many years turbine wheels have been manufactured by the same lost-wax method used for compressor impellers, page 13. However, turbine wheels cannot be made with the rubber-pattern process because the wax pattern must be ceramic-coated before the plaster mold is made. The wax pattern is dipped into a ceramic slurry as many times as needed to build up the desired coating thickness. The coating is allowed to harden between each dipping. Then the plaster mold is made. The ceramic protects the plaster so the mold will withstand the temperature of the molten alloy without breaking. After the mold hardens, it is heated to melt the wax, which is poured out. Molten alloy is poured into the cavity. Breaking the plaster reveals the turbine wheel.

Around 1955, some manufacturers began casting the turbine wheel and exducer as a single unit. This would have been done earlier but materials suitable for the high stress and temperatures had not been available at reasonable prices.

The next step was to braze, rather than bolt, the turbine wheel to the shaft. This method worked well but quality control was a problem. It was difficult to determine whether the braze was satisfactory until the turbocharger was run on an engine. If the braze was faulty, it was a little late to be finding out.

Since that time, several satisfactory bonding processes have been developed. These include inert-gas welding, resistance welding, electron-beam welding and friction welding.

Friction welding, Figure 2-19, has been used for turbine construction since about 1968. The turbine wheel is attached to a heavy flywheel and rotated at high speed. The rough-turned shaft is held in a vise. Simultaneously, as the motor rotating the turbine wheel is turned off, the shaft is jammed against a stub on the center of the turbine wheel.

The tremendous heat generated in both the shaft and the turbine wheel welds them solidly together. The whole process takes only a few seconds. Once the process time is worked out for a given wheel and shaft, a good weld is achieved every time.

Another current method is to cast the shaft and turbine wheel in one piece. In the early days, turbochargers

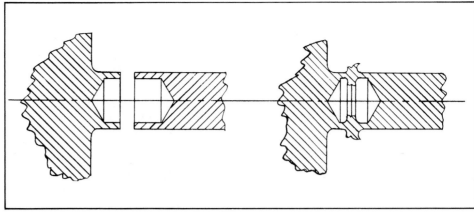

Figure 2-19—Before and after friction-welding shaft to turbine wheel

ran on bushings pressed into the housing. Contact between the bushings and shaft made it necessary to have extremely hard shaft journals to keep them from wearing out quickly.

With the floating-type bearings used on today's turbochargers, there is almost no bearing-to-journal contact. Journal hardness has dropped from about Rockwell C60 on the older shafts to about Rockwell C35 on shafts cast integrally with the turbine wheel. The cast shafts do not seem to wear out any faster than those with harder journals.

Several materials have been used in the past for turbine wheels, such as 19-9 DL, Stellite 31 and Stellite 151.

Today, nickel-based super alloys such as GMR 235 and Inco 713C seem to be most popular. These alloys will not corrode or melt under extreme conditions. In fact, turbines made from them will usually last until a nut, bolt or valve goes through them—or they have a bad rub on a housing due to a bearing failure.

Impeller Attachment—Some manufacturers press or shrink the compressor impeller on the shaft. Others prefer a looser fit.

If the turbocharger rotates counterclockwise—viewed from the compressor end—a right-hand nut on the compressor impeller will tend to tighten during operation.

If the turbocharger rotates clockwise, a right-hand nut will tend to loosen. Clockwise-rotation turbochargers with right-hand nuts must be assembled with the nut *very* tight—to the point of stretching the shaft as much as 0.006 inch when the nut is tightened.

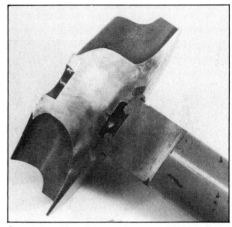

Section through shaft-to-turbine wheel friction-weld joint

BEARING HOUSING

The bearing housing between the compressor and the turbine is kind of a necessary evil to hold the whole thing together. Although bearing housings come in many shapes and sizes, they are basically the same. They have a bearing or bearings in the bore and an oil seal at each end.

Years ago, bearing housings had internal water passages to keep bearing temperature down because Babbitt was frequently used as a bearing material. If the engine was shut down from full load, heat soakback from the turbine wheel was enough to melt the Babbitt from the bearings unless there was a water jacket in the vicinity.

Turbochargers now use either aluminum or bronze bearings so heat soakback is not critical. Water jackets have been almost eliminated.

Unfortunately, the turbocharger is mounted in the hottest area of an al-

Figure 2-20—Cartridge-assembly cross section: This is a Roto-Master replacement unit.

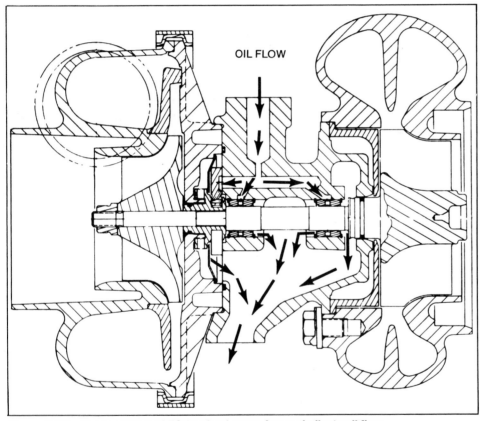

OIL FLOW

Figure 2-21—Cross section of TO4 turbocharger. Arrows indicate oil flow.

ready small engine compartment in some of the newer applications. One manufacturer uses a water-cooled bearing housing to prevent coking of the oil at the turbine end bearing. Some of the oil companies have been testing one-piece semifloating bearings as a low-cost way to achieve the same end. The one-piece bearing helps carry away excess heat from the turbine-end of the bearing.

Figure 2-20 shows a turbocharger cross section with the compressor and turbine housings removed. This assembly is usually referred to as a *cartridge* or *cartridge assembly*.

Bearing Lubrication—Pressurized oil enters the oil inlet and flows over and through the bearings. Clearance between the bearing and bearing housing and the bearing and the shaft journal is about the same. Oil flowing

between the bearing and the bearing housing tends to damp out vibration caused by imbalance in the turbine rotor. This imbalance, if great enough, would cause the journal to rub on the bearing. But the bearing has a slight bit of "give" due to the oil cushion, and contact between the journal and the bearing is practically eliminated. Thus, the life of the turbocharger is considerably extended.

In Figure 2-21, the cross-section of an AiResearch TO4 Turbocharger, some of the oil also enters the thrust bearing where it is ducted to the two thrust surfaces to lubricate the space between the thrust bearing and the thrust shoulders.

After the oil has passed through the bearing, it flows by gravity to the bottom of the bearing housing. From there it drains back to the engine crankcase through a hose or tube.

Oil enters the turbocharger from the engine at 30—80 psi. As it passes through the bearings, it is mixed with air and comes out looking like dirty whipped cream.

Because the oil flows from the turbocharger with no pressure, it is important to have the oil-drain line much larger than oil-inlet line. Equally important, this drain line can have no kinks or traps. This is covered in detail in Chapter 9, Lubrication.

Seals—When the turbocharger is doing its job, i.e., supercharging the engine, gas pressure behind the turbine wheel and the compressor impeller is much greater than the pressure inside the bearing housing. Seals must be placed between the bearing housing and the compressor impeller and turbine wheel.

Sealing the turbine end is relatively easy, despite the higher temperatures. Pressure in the turbine housing is always positive. The main job is to keep hot gases out of the bearing housing.

This is normally done by using a piston ring in a groove at the turbine end of the shaft, Figure 2-22. This ring fits snugly in the bearing housing and is also a very close fit between the walls of the ring groove. It does not rotate. The piston ring does a good job of keeping the hot gases out of the bearing housing.

Some turbochargers use a labyrinth-type seal on the turbine end. Here the hot gas is prevented from entering the bearing housing by a series of dams, Figure 2-23.

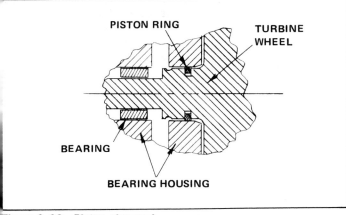

Figure 2-22—Piston-ring seal

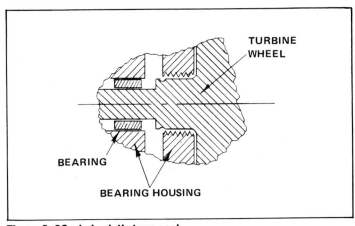

Figure 2-23—Labyrinth-type seal

The seal at the compressor end is a slightly different problem. On a diesel engine at idle or at very low power, the compressor produces practically no boost. Pressure drop through the air cleaner—even if it is in good shape—will cause a slight vacuum behind the compressor impeller. This vacuum will tend to suck oil from the bearing housing into the compressor housing.

Because the oil wets all the surfaces in the bearing housing, it is difficult to keep it away from the compressor end seal. Even a little oil leaking into the compressor housing makes things pretty messy in a hurry.

The compressor end seal in Figure 2-24 is designed for use on diesel engines and does a fairly good job. This seal has a piston ring and a small centrifugal pump built into the slinger to attempt to prevent oil from reaching the seal. This works fairly well on a diesel engine in which the compressor-intake vacuum is only a few inches of mercury.

However, when a carburetor for a gasoline engine is placed upstream of the compressor, it is possible to have almost total vacuum when the engine is running at high speed and the throttle is suddenly closed. A piston-ring type seal is not satisfactory under these conditions and will leak large quantities of oil into the intake manifold of the engine. However, it will work as long as the carburetor is downstream of the compressor.

Turbochargers designed for use with the carburetor upstream of the compressor usually have a mechanical face seal, Figure 2-25.

The Roto-Master line of turbochargers uses mechanical face seals on both gasoline and diesel

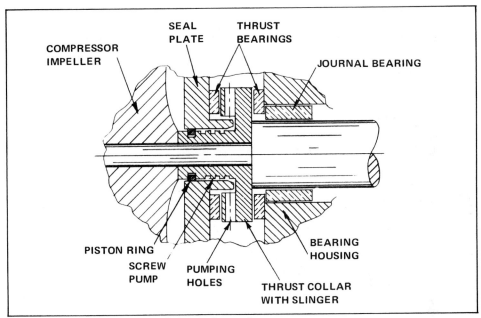

Figure 2-24—Compressor-end piston-ring seal

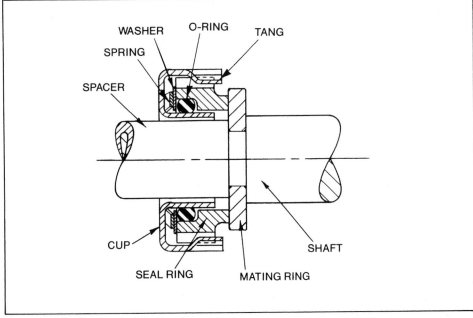

Figure 2-25—Mechanical face seal

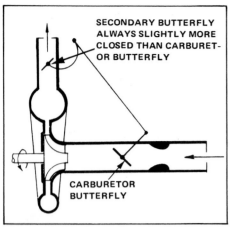

Figure 2-26—Secondary butterfly prevents vacuum at compressor inlet.

models. Other models usually have mechanical face seals only on turbochargers specifically designed for carbureted engines. Schwitzer has chosen to stay with the piston-ring seal for all applications.

Some carbureted applications have used a second butterfly downstream of the compressor. When properly connected to the carburetor or injector-butterfly, the second butterfly prevents high vacuum in the compressor housing, Figure 2-26.

Floating Bearings—As mentioned, the early automotive-type turbochargers had bearings pressed into the bear-

ing housing. These were similar to low-speed bushings, like a camshaft bushing. This type of bearing becomes unstable when used with a lightly loaded, high-speed shaft.

At extremely high speeds, the shaft starts to orbit around the center of the bearing bore, Figure 2-27. Movement is in the same direction of rotation as the shaft but somewhere below half the shaft speed. This unstable condition, known as *oil whirl* or *oil whip,* usually occurs near the natural frequency of the shaft. This is often referred to as the *critical speed.*

The motion becomes so violent that if the end of the rotor is observed on a test stand, it will suddenly blur due to the violent excursion of the shaft center away from the bearing center. It will sometimes appear to move as much as 1/16 inch. This seems impossible because the clearance between the journals and the bearing may only be 0.003—0.004 inch.

Once this has occurred for any length of time, examination of the bearing and journals will show severe wear on both, indicating the center of the shaft was actually moving far more than was allowed by the clearance.

Bearings have been designed with many different kinds of grooves and wedges to stabilize the shaft and prevent oil whirl. Nothing worked very well until floating bearings were adopted. As mentioned earlier, floating bearings have about the same clearance on the outside as on the inside, Figure 2-28.

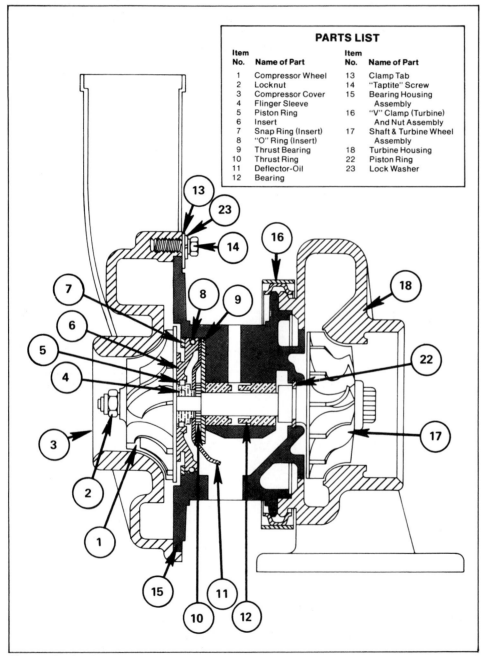

PARTS LIST

Item No.	Name of Part	Item No.	Name of Part
1	Compressor Wheel	13	Clamp Tab
2	Locknut	14	"Taptite" Screw
3	Compressor Cover	15	Bearing Housing Assembly
4	Flinger Sleeve		
5	Piston Ring	16	"V" Clamp (Turbine) And Nut Assembly
6	Insert		
7	Snap Ring (Insert)	17	Shaft & Turbine Wheel Assembly
8	"O" Ring (Insert)		
9	Thrust Bearing	18	Turbine Housing
10	Thrust Ring	22	Piston Ring
11	Deflector-Oil	23	Lock Washer
12	Bearing		

Figure 2-28—Cross section of Schwitzer 3LD turbocharger. Drawing courtesy of Schwitzer.

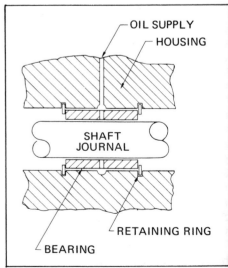

Figure 2-27—Floating bearing

24

Without putting any special grooves in the floating bearing, it does a great job of damping vibrations caused by rotor imbalance. And, it is almost impossible to determine the critical speed of the rotor because there is practically no difference in the vibration level of the turbocharger when running at the critical speed.

When turbochargers with fixed bearings were run on a test stand equipped with a vibration meter, 0.001-inch displacement was considered normal. Since the introduction of floating bearings, the vibration level has dropped so low that the limits of acceptability are now about 0.0002-inch peak-to-peak displacement.

Because these bearings are loose in the housing, they are free to rotate. Axial movement is prevented either by a retaining ring or a shoulder in the bearing-housing bore. Bearing rotation causes two problems—possible oil starvation and bearing-housing wear.

The bearing rotates at approximately one third the shaft speed, so when the shaft is turning 100,000 rpm, the bearing turns about 30,000 rpm. In spite of the fact that the bearing may only have a wall thickness of about 1/8 inch, holes through the bearing used to supply oil to its inner surface will act as a centrifugal pump. At 30,000 rpm the pressure differential between the inner surface of the bearing and the outer surface of the bearing may be enough to overcome the oil pressure applied from the engine. If this happens, no oil will flow through the bearing to the journal and rapid failure will occur.

Ways have been devised to prevent oil starvation. The Schwitzer Model 3LD turbocharger, Figure 2-28, uses a thinner wall in the bearing where the oil enters. This minimizes the pumping action. Holes in this area are also relatively large compared to the wall thickness, making it a very inefficient pump.

In Caterpillar Patent 3,058,787, E. R. Bernson attacked the problem in a different way. The oil passage that supplies the bearing has a wedge-shaped groove. This tends to increase the oil pressure as the oil is rotated from the oil-supply hole to the hole in the bearing.

The other problem with full-floating bearings—bearing-housing wear due to the rotating bearing—is partially

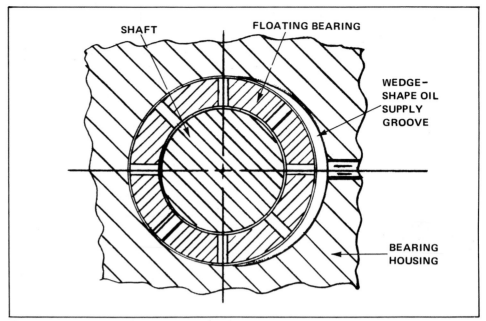

Figure 2-29—Bernson patent

solved by making the bearing housing from cast iron. Some bearing housings using full-floating bearings have been made from aluminum, but they were equipped with a hardened-steel sleeve in the bore. The cost of manufacturing and assembling a steel sleeve makes the use of an aluminum bearing housing prohibitive.

Semi-floating Bearings—When floating bearings first came into general use in turbochargers, it was assumed they had to rotate, so all the units were made in this manner.

Because oil flowing around the outside of the bearing does the damping, rather than the rotation of the bearing, the semi-floating bearing, Rajay Patent 3,043,636 (Hugh MacInnes) is used by some manufacturers. The bearing is kept from rotating by pinning the flange on the end of the bearing. Or, the flange can be matched to an irregularly shaped pocket in the bearing housing or the oil-seal plate, Figure 2-30. The flange prevents the bearing from rotating and also prevents it from moving axially. The wall thickness at the end of the bearing can be used for a thrust surface. Oil that passes through the bearing then passes between the shoulders on the shaft and the ends of the bearing to lubricate the thrust surfaces.

Because the semi-floating bearing does not rotate, it is possible to oper-

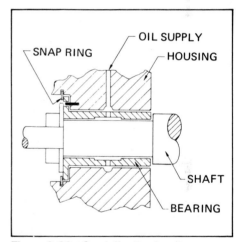

Figure 2-30—Semi-floating bearing

ate turbochargers of this type with very little oil pressure. Extremely low oil pressure is not recommended, but a turbocharger that will operate under adverse conditions has certain advantages. This type of bearing works very well in an aluminum bearing housing because it does not need a hard surface on the bore of the bearing housing.

SUMMARY

Small turbochargers are all basically simple, rugged devices. When treated properly, they will last as long as an engine. However, a turbocharger not treated properly may only last ten seconds.

3 Choosing the Engine

GTP Corvette is powered by this turbocharged 209-CID 90°-V6 Chevy. Engine must be durable, have a broad power range and good throttle response necessary for long-distance road racing. Photos by Tom Monroe.

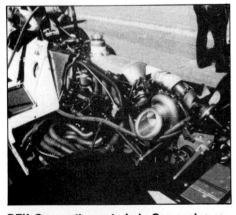

DFX Cosworth meets Indy Car-engine requirements for turbocharging: power, durability, weight, mileage and size. Although your application may not be as sophisticated, you'll have to consider some of these points. Photo by Tom Monroe.

One of the first questions usually asked by a person interested in turbocharging is, "Can I turbocharge my engine?" The answer is, "Yes, regardless of the make or age of the engine."

The next question is usually, "How much power can I get from it?" The answer to this one—"As much as the engine will stand."

And this one is usually followed by, "At what rpm will I start to get boost?" The answer here often startles the unknowledgeable: "At whatever rpm you want."

The fact is, any automobile or motorcycle engine from about 30 CID (500cc) or larger can be turbocharged with presently available turbochargers. Results will depend on how strong the engine is and how well the installation is made. There isn't an engine built today that cannot be turbocharged beyond its physical capabilities.

Turbochargers have already been *staged*—two or more turbochargers used in series—to produce intake manifold pressures above 200 psig. *This pressure can be increased without exceeding the capabilities of existing turbochargers.*

Where and when maximum boost is achieved is covered in Chapter 4, Choosing the Turbocharger. This should be done after it is decided what the use of the vehicle is going to be and how much time and money is to be invested.

Because any engine can be

Pinto 2.0-liter four responds well to turbocharging and is very durable. Photo courtesy of Petersen Publishing.

258-CID small-block Chevy powered the Holmes, Kugel & McGinnis '29 Model A roadster to record 245.804-mph average at Bonneville. Two AiResearch TO4 turbos pressurize Hilborn fuel injection. Photo courtesy of Gray Baskerville/Hotrod Magazine.

Datsun fans in the United States will not be familiar with this Janspeed kit for a L26 and L28 6-cylinder Datsun engine. Waste gate is below turbine inlet. Compare this simple installation, which uses a non-crossflow head, to the Alfa Romeo installation shown at right.

Crossovers must be used on crossflow heads: 2-liter Alfa Romeo Alfetta turbocharged by Mathwall for Bell and Colvill. Turbocharger blows through SPICA fuel injection. Alfa goes 0-60 in 8.1 sec.

turbocharged, it is not necessary to narrow the choice down to one or two models. It is certainly easier to turbocharge an engine for which a bolt-on type kit is available. However, the kit may not be designed to produce maximum horsepower at the speed you want for your application. Most kits are designed for street use and must be modified for such things as drag racing and/or track racing.

For someone turbocharging an engine for the first time, an in-line four- or six-cylinder engine presents the fewest problems. Those with non-crossflow heads—intake and exhaust manifolds on the same side—simplify the installation still further.

When turbocharging an opposed or V engine, there's a problem in ducting half of the exhaust gases to the other side of the engine. Most V-6 engines are too small to consider the use of two turbochargers, but two small or medium-size turbochargers work very well on a V-8.

A few years ago fours and sixes were more desirable than the eights cost-wise, because they usually had a low compression ratio. The eights almost invariably had high compression ratios up to 10:1. This has all changed because a low compression ratio is one of the ways the engine manufacturers have reduced emissions. The effect of turbochargers on engine emissions is covered in Chapter 19.

When engines had high compression ratios they were designed to run well on 100-octane fuel. This limited the amount of power that could be gained easily with the addition of a

turbocharger. Ak Miller's rule of thumb says, "The octane requirement of an engine goes up one point with each pound of boost from the turbocharger." For example, if an engine with 8:1 compression ratio runs well on 90-octane gasoline, it will require 98-octane to run with eight pounds of boost and no detonation. Anything above this may require water injection.

Because just about all production engines available since the mid-70s have low compression ratios, the choice of engines that can be highly turbocharged without any major modification is unlimited.

Flathead engines disappeared with the advent of high-compression overhead-valve engines. This was probably because the extra volume of the flathead combustion chamber does not lend itself to high compression ratios. In addition, most of these engines had crankshafts with only one main bearing for every two throws—not recommended for big horsepower increases.

Turbochargers tend to be equalizers with respect to the breathing ability of the engine. It would be interesting to see a comparison between a turbocharged flathead engine and an overhead-valve engine of the same displacement. It would not be surprising if the flathead engine produced about the same horsepower per cubic inch with the same boost pressure.

Natural gas is often used to fuel engines that power irrigation pumps. When turbocharged, one of these engines can replace two naturally aspirated engines. This 542-CID International Harvester engine was converted for natural gas and dual turbos by Ak Miller Enterprises.

Turbocharging an inline engine with a crossflow head adds complications. Spearco chose to blow through the carburetor to minimize distance from carburetor to intake ports. Turbine and waste-gate outlets of IHI RHO5 unit on this 1600cc Ford Escort are joined for a simpler exhaust.

Not many kit manufacturers have been brave enough to tackle rotary engines. Kas Kastner of Arkay went all the way with cast manifolds and adapters. Engine produces 202 BHP at 7000 rpm with 7-ps boost.

Choosing between an air-cooled and a liquid-cooled engine is not much of a problem. Most engines are liquid-cooled. But the air-cooled VW engine takes very well to turbocharging. A 100% power increase is possible on the VW with no other engine modifications.

The biggest single problem with turbocharging air-cooled engines is keeping the cylinder heads cool. If the heads get too hot, detonation will destroy the engine in a matter of seconds. It is advisable to use a head-temperature indicator with some kind of a buzzer or light to provide a warning before detonation occurs. Corvair Spyders had such a warning device. Another possibility is using water-cooled cylinder heads—as Porsche did on some of their racing engines.

The Wankel engine has been turbocharged with good and bad results, the same as with reciprocating engines. Detonation should be avoided on a Wankel because it may destroy the apex seals.

TRANSMISSIONS

Manual—With a manual transmission, it doesn't take long to find out that the turbocharged engine requires different shift points than the naturally aspirated engine. Fortunately, manual transmissions don't seem to care one way or the other. Modern three- and four-speed

Art Nolte of Scottsdale Automotive Specialists ran this turbocharged AA dragster. Triple turbos are AiResearch TO4's with V-trim. Engine is a 481-CID Rodeck big-block Chevrolet with Delta crank, Childs & Albert rods, Aries pistons and Howard cam. Hilborn injectors feed straight methanol. Speed of 204 mph and elapsed times of 6.80 seconds were recorded in 1981. Car was handicapped by a 200-pound weight penalty levied against turbocharged cars.

manual transmissions have enjoyed many years of development. They seem to stand up well under the increased torque produced by a turbocharged engine.

Automatic—Automatic transmissions are another story. Gears and bearings are usually strong enough to put up with considerably more horsepower. The shifting mechanism, however, is designed to accept the torque of the naturally aspirated engine. Internal clutches and brake bands may not have enough surface area to work without slipping when extra torque is available at the shift points.

A symptom of this weakness arises during full-throttle acceleration where slippage occurs at each gear change. Slippage will get worse with use and time. Eventually, the transmission will not even transmit engine torque at road load.

One of the ways to prevent this is to install a heavy-duty shift-programming kit to increase clutch and band clamping force, thereby increasing torque capacity. A full-throttle shift then becomes a real boot in the tail. This can sometimes be minimized by changing the shift points, which may also be accomplished by the kit or with other modifications.

With some modern-day cars, the automatic transmission has been designed so there is little, if any, reserve

Turbocharged 120-CID Ford Pinto engine pushed this streamliner to an E Class record of 203.84 MPH at the Bonneville National Speed Trials. Owned by Bennett & Ruiz of Santa Barbara, California, the engine relies on an AiResearch turbocharger for its outstanding performance.

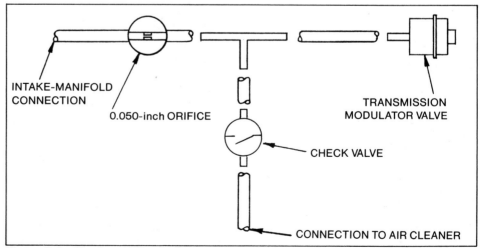

INTAKE-MANIFOLD CONNECTION

0.050-inch ORIFICE

TRANSMISSION MODULATOR VALVE

CHECK VALVE

CONNECTION TO AIR CLEANER

Figure 3-1—Simple addition to the vacuum-modulator sensing line usually solves shifting problems on vacuum-modulated automatic transmissions.

B&M Modulink replaces vacuum modulator on turbocharged applications. Vacuum line is replaced by cable that uses throttle position to indicate engine load.

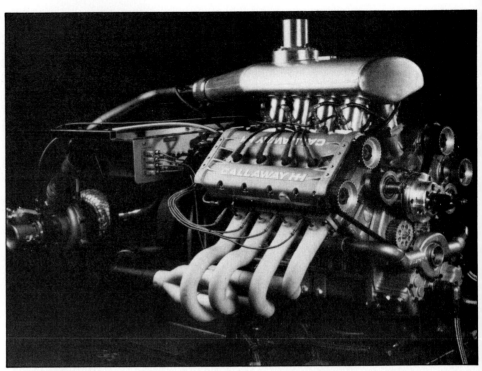

Ideal situation; develop engine *and* turbocharger as a package. Few can afford this luxury. Dubbed *Callaway HH V8*, Reeves Callaway is developing his engine with this Warner-Ishi-turbo'd 160-CID Indy Car version. Engine testing is done on computerized Superflow dynamometer. Photo by C. E. Green.

capacity. In such cases it may be necessary to install an automatic with more capacity.

The naturally aspirated engine/transmission combination is designed to shift, if possible, when the torque delivered by the transmission before and after the shift is approximately equal. When this is done, the shift point is almost imperceptible.

The turbocharged engine, however, is a different story. Assume the engine has revved up to 4500 rpm before the shift and the turbocharger is putting out 7-psi boost. If engine speed drops to 3000 rpm the turbocharger will still be producing at least 7-psi boost. This boost at low rpm really produces a lot of torque. Consequently, the transmission shifting mechanism does things it was not designed to do.

When and how the transmission shifts is controlled by the transmission valve body. It operates on the basis of signals received from throttle-linkage position, intake-manifold vacuum and road speed.

One or perhaps all of these signal inputs may have to be modified when the engine is turbocharged. If the changes are done by a transmission expert, it is possible to have a very responsive setup with outstanding performance. This assumes that the transmission is capable of handling the extra power.

Many automatic transmissions have what is known as a *vacuum modulator*. This device uses intake-manifold vacuum to signal the transmission about engine load. Boost pressure confuses transmission operation and may rupture the diaphragm in the modulator.

To protect the modulator, install a check valve on a T in the line to the vacuum modulator, Figure 3-1. The check valve will bleed off pressure when the turbo is supplying boost. When there is manifold vacuum, the valve will close and allow normal modulator operation.

The check-valve opening can be quite small—a 1/16-inch to 3/32-inch bleed hole will be adequate. The outlet should be vented to the air cleaner or carburetor inlet because it will contain the fuel/air mixture that is present in the intake manifold.

B & M Transmission has a mechanical device called a *Modulink*. This provides proper throttle-position sensing for vacuum-modulated transmissions used with turbocharged engines. The B & M Modulink replaces the vacuum modulator with a cable that connects to the carburetor throttle lever. The Modulink provides adjustable throttle-position sensing for consistent shifting.

SUMMARY

If it runs, it can probably be turbocharged. But as with modifying naturally aspirated engines, some engines become considerably more popular than others. This is simply because the parts are available and some engines can stand a lot more overload than others.

4 Choosing the Turbocharger

Using the correct turbocharger on an engine creates more power than by any other means. However, putting any turbocharger on an engine does not *guarantee* good results. If a turbocharger is a good match for a 200-CID engine at 4000 rpm, it will probably not be a good match for a 400-CID engine at 4000 rpm.

Some people have the notion that by running the turbocharger twice as fast, it will work on an engine twice as large. It just isn't so!

A look at any of the compressor maps in this chapter or the appendix will show you that the turbocharger speed lines are relatively horizontal in the operating range. *Increasing the speed of the turbocharger increases the pressure ratio, not the flow.* Turbocharger speed is varied by changing the turbine-nozzle size. If the compressor is not sized correctly to the engine, changing the turbine nozzle will not help.

For those willing to go through the process of choosing the correct size turbocharger, I explain the process in detail in the next few pages.

However, it is apparent from talking to many enthusiasts that they don't feel confident in doing it by this method. I've included a selection of turbochargers to be used on most engines and applications. It's at the end of this chapter.

Engine Application—Don Hubbard suggests classifying turbocharged engines by application, Table 2. This is because the user will want maximum torque and horsepower at an rpm range dependent on the application, Table 2.

Table 2 is very useful in matching the correct compressor to the engine. It takes into account whether the engine is to be run at steady power as in an offshore-racing boat or on-and-off as in a road-racing machine. This is important when choosing the turbine-nozzle size or deciding whether or not to use a wastegate or other controls. Table 2 points out the engine-speed range necessary for various applications and the power-burst duration. The engine speed selected, plus the required intake-manifold

BMW engine on dyno at McLaren Engines. Extensive testing is necessary to match all engine-control systems to obtain optimum power output. Matching starts with sizing turbocharger to engine, regardless of application or degree of sophistication.

Turbocharged 1497cc (91-CID) Escort/Lynx engine powered Don DeBring's streamliner to 269.196 mph at the Bonneville Salt Flats. Turbo is sized for maximum high-rpm power. Throttle response and low-end torque needn't be considered for such an application. Photo by Tom Monroe.

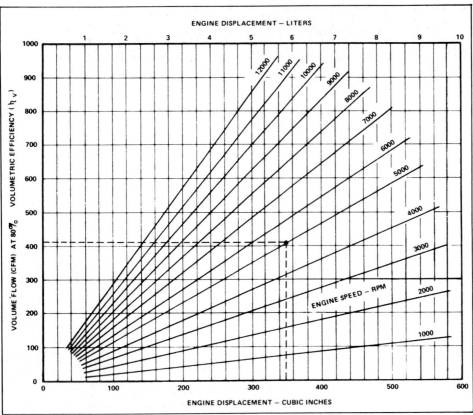

Figure 4-1—Naturally aspirated volume flow for 4-stroke engines. Dotted line is for a 350-CID engine at 5000 rpm.

Figure 4-2—Density ratio versus pressure ratio

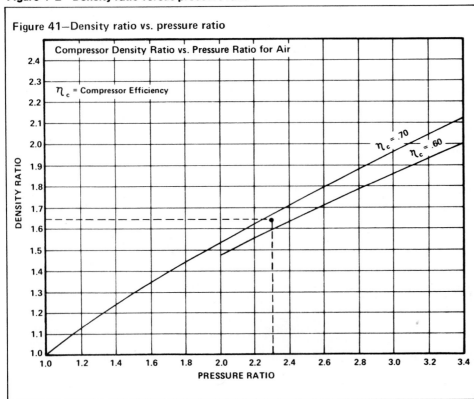

pressure, are the key factors in choosing the compressor for the engine.

Once you've decided the engine application, a number of calculations must be made. Figures 4-1 and 4-2 allow you to calculate volume flow for the turbocharger at peak conditions.

For example, from Table 2—Application 2, start with a 350-CID, 7:1 compression ratio engine running at 18-psi boost. You want a turbocharger that will meet these conditions at 5000 rpm.

Referring to Figure 4-1, start at 350 CID as shown on the dotted line and draw a line vertically to intersect the 5000-rpm line. At this point, draw a horizontal line left to intersect the volume-flow scale. In this case the engine will flow 410 cfm. This chart has been compensated for a volumetric efficiency of 80%, a good figure for average engines.

Figure 4-2 is compressor density ratio vs. pressure ratio for air. Because the turbocharger squeezes the air down considerably from ambient conditions, this chart is necessary to calculate the amount of air entering the compressor.

All centrifugal-compressor maps are based on inlet conditions. To use the chart, it is first necessary to convert boost pressure to pressure ratio.

Assuming the engine is to be run at approximately 1000-feet altitude where the ambient pressure is 14.3 psia (Appendix, page 144), the pressure ratio is computed:

Pressure Ratio =

$$\frac{\text{Ambient Pressure + Boost Pressure}}{\text{Ambient Pressure}} =$$

$$\frac{14.3 + 18}{14.3} = 2.26$$

Most of the running done on turbochargers available today will be between 60% to 70% compressor efficiency, and only these two lines are shown on the chart.

In the example, I split the difference and took it to about 65%. Draw a vertical line from 2.26 pressure ratio to one of the efficiency lines. From here draw a horizontal line to the density ratio scale on the left. The density ratio is 1.62.

Under the given conditions in the example, the engine will flow 1.62 times more air with the turbocharger. Go back to the naturally aspirated volume-flow chart where we obtained 410 cfm. Multiply 410 by 1.62 density ratio to get 664 cfm at 2.26 pressure

TABLE 2
TURBOCHARGER APPLICATIONS

	Category	Engine Speed Range for Maximum Torque	Duration of Power Burst	Maximum Boost Pressure psi	Fuel
1	Street machine	As wide as possible	10 seconds maximum	10	Gasoline or propane
	Truck or bus	Medium to high	Continuous	10	
	Sport fishing boat	High	Continuous	10	
2	Road racer	As wide as possible	Some short, some long	20	Gasoline or propane
	Drag racer	Medium to high	10 seconds maximum	20	
	Short-course racing boat	High	Off and on but almost continuous	20	
	Drag racing boat	High	10 seconds maximum	20	
3	Same as above			30	Gasoline with water-alcohol injection
4	Oval-track racer	Medium to high	Almost continuous	45	Methanol
	Long-course racing boat	High	Continuous	45	
	Drag-racing boat	High	10 seconds maximum	45	
5	Tractor-pulling contest	Medium to high	2 minutes	40	Diesel
6	Tractor-pulling contest	Medium to high	2 minutes	110	Diesel

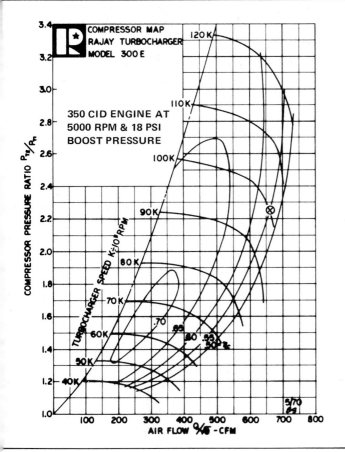

Figure 4-3—Rajay Model 300E compressor map

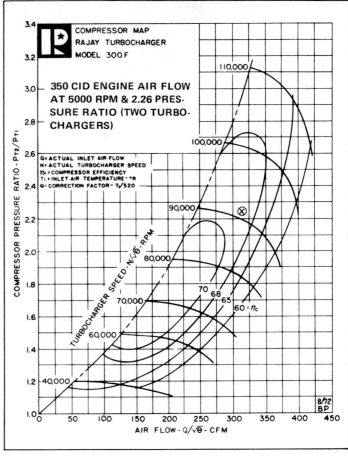

Figure 4-4—Rajay Model 300F compressor map

33

Janspeed-turbocharged Mini Metro blows through 40 DCOE Dell'Orto carburetor.

Installing a turbocharger in a tight engine compartment can be difficult. Such is the case with a Buick Skyhawk. Doug Roe managed it quite well in this Buick V6 kit offered by Kenne-Bell. An SU or Dell'Orto is used.

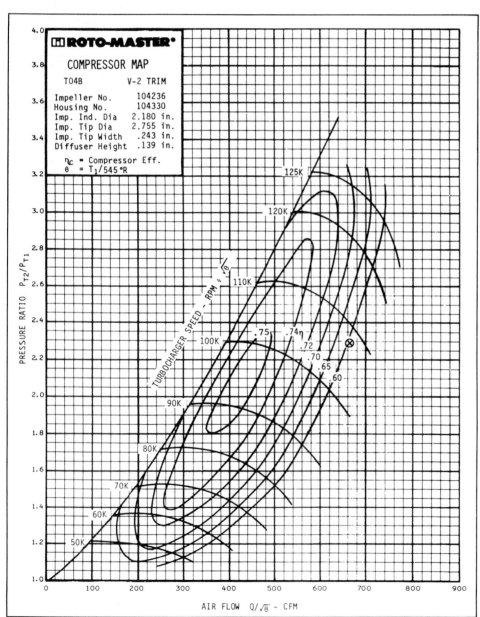

Figure 4-5—Roto-Master TO4B V-2 trim compressor map

ratio. You can now start looking at compressor maps for specific turbochargers.

Plotting this point on the Rajay Model 300E map, Figure 4-3, puts it off to the right at about 55% efficiency. Any point below 60% compressor efficiency should be avoided, so look elsewhere.

However, if two turbochargers were used on this engine with 332 cfm each at 2.26 pressure ratio, this point looks very good on the Rajay 300F map, Figure 4-4. Here it falls in an area above 68% efficiency.

If the Roto-Master TO4B Model is used, the V-2 *trim* is borderline for a single installation, Figure 4-5. An S-4 *trim*, Figure 4-6, will work well on a dual-turbo installation. An R-11 Roto-Master, Figure 4-7, will be sufficient for a single-turbo installation.

TRIM
As shown in Figure 2-18, compressor impellers and turbine wheels are trimmed to various contours (trims) by machining. These *trims* determine the flow range of the compressor or turbine. Each trim has an identifying number or letter assigned by the manufacturer.

This points out that more than one model or make of turbocharger will frequently do a good job for a specific application. Your choice is not limited to one size. You are not forced to compromise and choose a turbocharger which is not a good match for the engine.

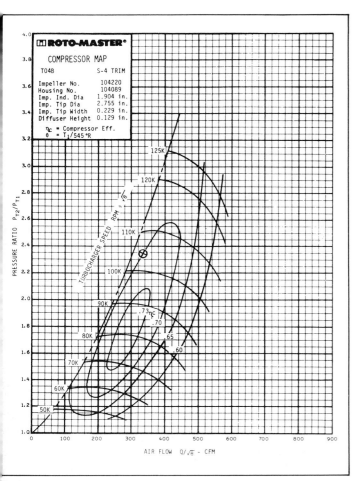

Figure 4-6—Roto-Master TO4B S-4 trim compressor map

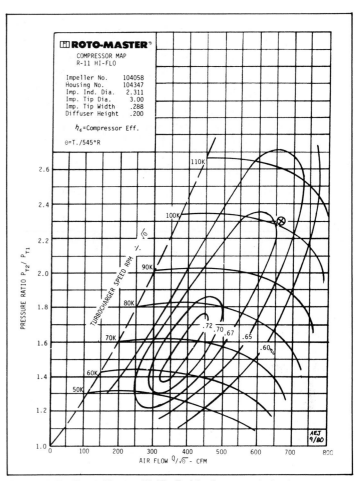

Figure 4-7—Roto-Master Hi-Flo R-11 trim compressor map

MULTIPLE INSTALLATION

Another decision to be made is whether to use one or two turbochargers. If the engine is a small in-line four, it is easy to decide on one turbocharger. Even an in-line six looks good with one turbocharger unless the desired output is very high.

When it comes to a V or opposed engine, other factors must be considered. If one turbo is to be used, you have to pipe the exhaust gases from one side to the other, or at least join the two exhaust manifolds. Because the exhaust pipes get considerably hotter than the rest of the engine, they expand more. Unless some sort of flexibility is built in, they will eventually crack and leak exhaust pressure. This is discussed in Chapter 15, page 108.

Using two smaller units eliminates this exhaust-system problem because the turbocharger system needs to be connected only on the compressor side. It is often easier to fit two small turbochargers under the hood than one large one.

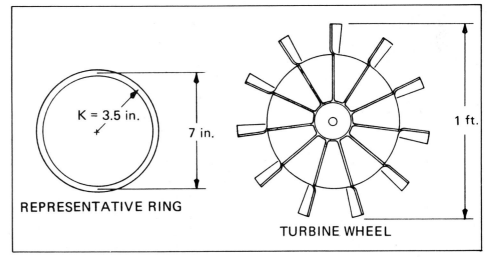

Figure 4-8—Turbine wheel and representative ring have same moment of inertia.

Reducing Turbo Lag—Besides installation problems, the inertia of the rotor should also be considered. One of the standard complaints about turbochargers is, "They lag behind the engine." Lag can be due to carburetion or ignition-timing problems, but reducing rotor inertia is a step in the right direction.

As soon as the throttle is opened, there is a sudden increase in exhaust-gas volume and the turbocharger ac-

Porsche was one of the first companies to use turbocharged engines successfully in road racing. Dual-ignition flat-12 from 917 (on palette) produced 1100 HP from 5.4 liters. Porsche 917 won the World Championship of Makes (now the World Endurance Championship) in 1970 and '71. Both the 917 engine and the 936 (in car) use twin turbos. This 2.6-liter flat-6 936 engine produced 620 HP.

celerates rapidly. Time required for the turbocharger to reach maximum speed is a function of overall turbocharger efficiency and the polar moment of inertia of the rotating group.

Moment of inertia is the resistance of a rotating body to a change in speed, represented by the letter I.

$$I = K^2 M$$

where K is the radius of gyration and M is the mass of the body.

Radius of gyration is the distance from the rotating axis to the point where all the mass of the body could be located to have the same I as the body itself.

In other words, a 12-inch-diameter turbine wheel might be represented by a ring with a diameter of 7 inches. In this case, K = 3.5 inches, Figure 4-8.

For good rotor acceleration it is essential to have the lowest possible moment of inertia. Turbine wheels are designed with a minimum of material near the outside diameter to reduce K. The moment is proportional to the *square* of K, so reducing K by 1/2 will reduce I to 1/4 of the previous value.

Because of this, it is often advantageous to use two small turbochargers rather than one large one.

As an example, a 3-inch-diameter turbine weighing 1 pound with a K of 1 inch will have a moment of inertia

$$I = K^2 M$$

$$= K^2 \frac{W}{G}$$

where G is the acceleration of gravity and W is the weight

$$I = 1 \times \frac{1}{386}$$

$$= 0.00259 \text{ in-lb-sec}^2$$

If one larger turbocharger is used instead of two small ones, it may have a turbine diameter of 3.5 inches and a rotor weight of about 1.5 pounds. Assuming a radius K of 1.25 inches, the amount of inertia will be:

$$I = K^2 \frac{W}{G}$$

$$= \frac{1.25^2 \times 1.5}{386}$$

$$= \frac{1.56 \times 1.5}{386}$$

$$I = 0.00607 \text{ in-lb-sec}^2$$

This moment of inertia is about 2.5 times greater than the smaller turbocharger.

Where turbocharger acceleration is important, I recommend you use two small units instead of one large one.

One of the toughest applications is that of the road racer. There are times when the car is making a turn with the throttle closed and the turbo running at low rpm. The throttle will then be opened wide and the engine will immediately demand, for example, 100 cfm of air.

As shown on the compressor map in Figure 4-9, this will happen with no boost. The turbocharger will accelerate immediately but won't catch up with the engine until steady-state conditions are reached. If the time between maximum acceleration and

shut-off is short, the engine may never reach steady-state conditions.

Other Applications—The fact that the turbocharger is not mechanically connected to the engine and must overcome inertia does have some advantages. Many applications, including drag racing, require shifting gears.

This time we will look at the path on the compressor map which represents this application, Figure 4-10.

The vehicle will be on the line with no boost and will start out similar to the previous illustration. When the shift is made, engine speed will drop off while turbocharger speed remains constant. Because the compressor puts out more boost at the lower flows, there will be an immediate increase in manifold pressure. This occurs at each shift point and gives the car a noticeable push.

Dotted lines parallel to the turbocharger speed lines are drawn to show how much boost increase will occur at each shift point. These curves are general and the actual amount will depend on the engine/turbocharger/gear combination.

TURBOCHARGER MATCHING

Sizing and matching a turbocharger for a short-duration application must take into account both the maximum boost desired and turbocharger acceleration. This is another reason vehicle application is so important in turbocharger selection.

A bus or truck engine must be able to live with the maximum boost

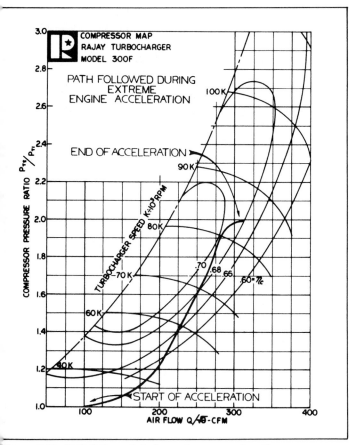

Figure 4-9—Path followed during extreme acceleration

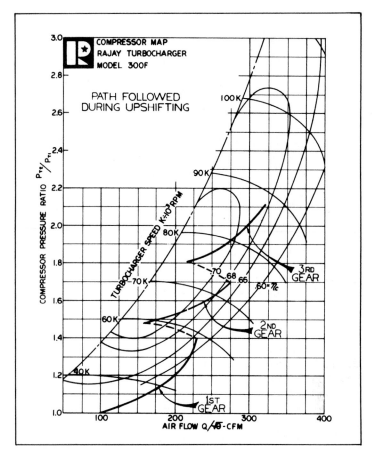

Figure 4-10—Path followed during upshifting

which will occur when climbing a long hill. A drag-race engine has to produce full power for only a few seconds. In this case, the turbocharger must be matched not only for a higher boost but for the acceleration lag as well.

There is no set rule on how to arrive at a perfect match. Each application must be matched either by "cut and try" or experience. This is where the tables at the end of this chapter will prove very useful. The fellows who supplied the information did it based on actual experience. The main thing is to know which way to go after some results have been obtained.

The first requirement is to find out how much air will flow through the engine at maximum speed. This leads to the question: "What is the maximum speed?"

In general, the camshaft design determines the engine speed at which peak torque is produced. It also establishes the engine rpm at which peak power is produced. The turbocharger should be matched to increase output in the rpm range from peak torque to peak power.

A turbocharged engine does not always require as much valve overlap as a naturally aspirated engine. Too much overlap can actually reduce the output by cooling down the exhaust gases. This slows down the turbine and therefore produces less boost.

As I mentioned, many turbocharger enthusiasts don't like going through a lot of calculations to determine which size unit to use for their particular applications. For those interested, the following example goes through the whole process. I used this kind of calculation to derive the charts at the end of the chapter.

Engine Requirements—First of all, we must start with some known quantities:

Engine Size	250 CID
Engine Type	Four-Stroke
Maximum Speed	5000 rpm
Maximum Boost Pressure	10 psig
Ambient Temperature	70F (530R)
Barometer	29.0 in. Hg

Start by converting the displacement to cubic feet.

$$\frac{250 \text{ cu in.}}{1728 \text{ cu in./cu ft}} = 0.145 \text{ cu ft}$$

Next calculate the ideal volume flow through the engine.

$$\frac{0.145 \text{ cu ft}}{\text{Revolution}} \times \frac{5000 \text{ rpm}}{2} = 362.5 \text{ cfm}$$

Speed is divided by 2 because this is a four-stroke engine. Therefore, it will flow 362.5 cfm if the volumetric efficiency (η_{vol}) is 100%.

Because of port restrictions and residual gases left in the combustion chamber, this engine will probably flow only about 80% of its theoretical capacity or 0.8. Actual capacity is:

362.5 cfm x 0.8 = 290 cfm

Air entering the intake manifold will not be at standard conditions (29.92 in. Hg and 60F), so it is necessary to compute the actual density to determine air flow through the turbocharger compressor. All compressors are rated on inlet flow so the outlet flow (engine flow) must be multiplied by the density ratio across the compressor to determine the inlet flow.

It would be nice if we could just multiply the outlet flow by the pressure ratio, but the temperature rise

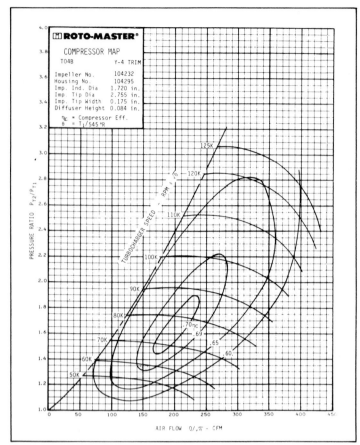

Roto-Master TO4B Y-4 trim compressor map

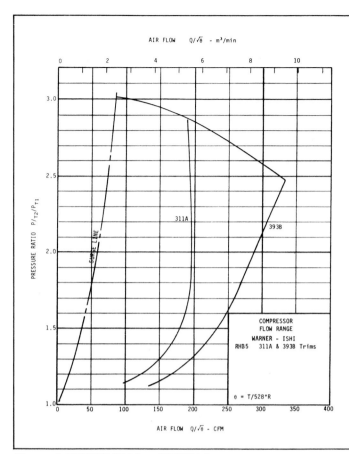

Warner-Ishi RHB5 311A and 393B trim compressor map

which accompanies the pressure rise makes the gas less dense than it would be if the temperature remained constant. This is discussed in detail in Chapter 11, Intercooling.

Assuming 65% compressor efficiency, calculate the density of the intake-manifold air

$$\text{Pressure Ratio} = \frac{\text{Manifold Pressure Abs}}{\text{Inlet Pressure Abs}}$$

$$= \frac{10\,\text{psig} + \text{Barometer}}{\text{Barometer}}$$

$$= \frac{(10 \times 2.03) + 29.0}{29.0}$$

$$= 1.7$$

Multiplying by 2.03 converts psig to in.Hg.
From Table I, Page 16
Where:

$$r = 1.7$$
$$Y = 0.162$$

Ideal Temperature Rise ΔT_{ideal}

$$= Y \times T_1$$

Because inlet temperature is 70F,

$$\Delta T_{ideal} = 0.162 \times (70F + 460F)$$

It is necessary to add 460F to the temperature because this calculation must

be in absolute temperature, i.e., degrees Rankine.

$$\Delta T_{ideal} = 85.9F$$

Actual Temperature Rise,

$$\Delta T_{actual} = \frac{\Delta T_{ideal}}{\text{Compressor Efficiency}}$$

$$\Delta T_{actual} = \frac{85.9}{0.65}$$

$$\Delta T_{actual} = 132$$

Intake-Manifold Temperature
= Compressor-Inlet Temperature + ΔT_{actual}

$$= 70 + 132$$
$$= 202F$$

Air in the intake manifold will not actually get this hot because fuel has a cooling effect. There will also be some heat loss through the ducting.

Comparing this to the air at the compressor inlet will result in the density ratio.

Density Ratio

$$= \frac{\text{Inlet Temperature}}{\text{Outlet Temperature}} \times \frac{\text{Outlet Pressure}}{\text{Inlet Pressure}}$$

$$= \frac{70 + 460}{202 + 460} \times \frac{20.3 + 29.0}{29.0}$$

$$= \frac{520}{662} \times \frac{49.3}{29.0}$$

Density ratio = 1.36

Under these conditions, the actual compressor inlet flow is:

Compressor-Inlet Flow

= Compressor-Outlet Flow (Engine Flow) x Density Ratio

= 290 cfm x 1.36
= 394.4 cfm

If you are using a compressor ma scaled in lb/min rather than cfm it i necessary to convert by multiplyin by 0.069 to obtain the correct flow be cause most lb/min compressor map are created by running the turbocharg er compressor at 85F and 28.4-in. H pressure.

For convenience, all the maps i this book are scaled in cubic feet pe minute.

The above calculations show 394. cfm flowing through the compresso at a 1.7:1 pressure ratio. It's best t have this point on the righ (high-flow) side of the map because i will occur at maximum engine rpm.

Try plotting this point on severa maps to determine which is best. N doubt any one of several turbos wi do the job, so you'll have a choice

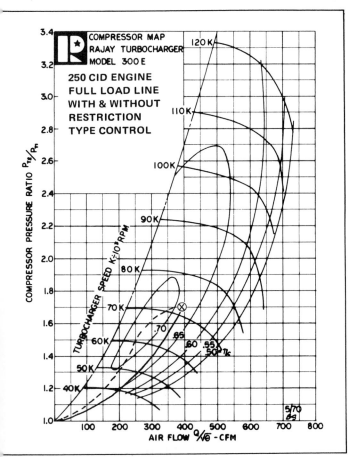

Figure 4-11—250-CID engine on compressor map

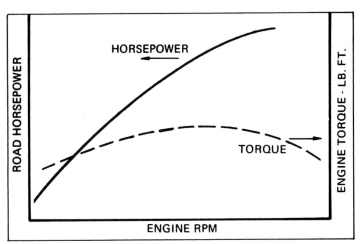

Figure 4-12—Road horsepower

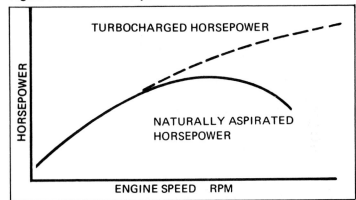

Figure 4-13—Turbo matched for high rpm

The Roto-Master T04B V-2 trim looks very good. It will produce this flow and pressure ratio at about 73% compressor efficiency.

A Rajay 300E flow will also work well with air flow to spare.

This is not an attempt to limit the user to these makes of turbochargers. It is just difficult to match an engine and turbocharger without a compressor map or previous experience.

ADJUSTING THE POWER CURVE

Turbine-Housing Selection—It is best to start with a large turbine-housing A/R or nozzle area. A too-small housing might overboost on the first run and destroy the engine. If the housing is too large, the boost will be low, but at least the engine will not be harmed.

An engine with the largest-possible carburetor and lowest-possible exhaust restriction has a wide-open-throttle acceleration line like the solid line on Figure 4-11. This leaves something to be desired at the mid range.

If a smaller turbine-housing A/R or nozzle area is used, the engine will be overboosted at maximum rpm. Overboosting can be avoided by using a control device such as a wastegate. Controls are covered in Chapter 10.

It is possible to have a more desirable wide-open-throttle operating line, as shown by a dotted line in Figure 4-11. It is possible to achieve this condition without using a movable control. Any one or a combination of the methods described here will improve the mid-range torque without overboosting at maximum engine speed.

Carburetor Restriction—Assume a carburetor that creates a pressure drop of about 4 in. Hg at the 5000-rpm point is used. The depression at the compressor inlet will make it necessary to use a smaller turbine housing of perhaps A/R 0.8 to achieve the original boost level of 10 psi. This carburetor will have a lower pressure drop at the lower speeds and the smaller turbine housing will still deliver adequate boost. This arrangement will follow a speed line similar

to the dotted line on Figure 4-11. The use of a smaller A/R turbine housing will cause more back pressure on the engine, which might raise exhaust-valve temperature. This is only critical when the engine is used for steady-state high output.

Exhaust Restriction—A similar result may be obtained by restricting the turbine exhaust, causing a back pressure on the turbine of about 4 in. Hg at maximum speed. Although a turbocharged engine is relatively quiet without the use of a muffler, this will result in an even quieter-running engine.

Valve Timing and Duration—A camshaft profile that favors a certain engine speed naturally aspirated will have the same tendency turbocharged. Our theoretical engine will reach peak boost at 5000 rpm. But if the cam and valve train are designed to favor 4000 rpm, it is possible to match the turbine housing so the boost again will follow the dotted line on the compressor map, Figure 4-11.

Pressure Differential—One of the measures of overall turbocharger per-

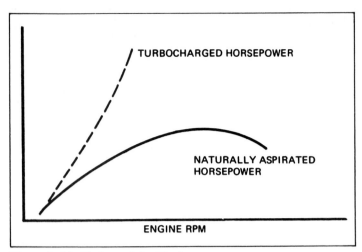

Figure 4-14—Turbo matched for low rpm

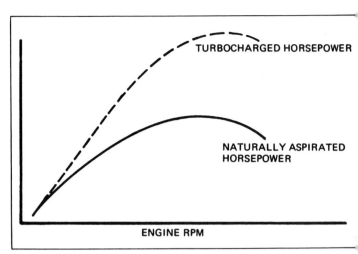

Figure 4-15—Turbo matched for broad speed range

formance is the amount of *positive differential* across the engine. This differential occurs when intake-manifold pressure is higher than exhaust-manifold pressure. It is extremely beneficial to the engine.

Almost all gasoline and liquified petroleum gas (LPG) engines have either carburetors or port fuel injection. Either system works well with a moderately turbocharged engine.

The combination of intake and exhaust restrictions and the fact that most turbocharging takes place only under extreme acceleration conditions causes the exhaust-manifold pressure to be higher than intake-manifold pressure at all times.

Care should be taken when using a high-overlap cam and an intake and exhaust system with low restriction. Under steady-state, high-pressure turbocharging, it is possible to blow a mixture of air and fuel through the combustion chamber into the exhaust manifold. This mixture can be ignited by the hot exhaust manifold and cause either of two things—or both! It may explode intermittently or burn continuously. Either causes problems.

If the mixture explodes, it may cause a power loss or damage the engine and turbocharger. If it burns continuously, it can raise turbine-inlet temperatures to 2000F. This will definitely shorten the life of the turbo and exhaust valves.

If valve overlap can be reduced without losing power, the tendency to blow raw fuel into the exhaust manifold will also be reduced.

Reducing the size of the turbine wheel to one of lower capacity may

also help if it is large enough to handle the gas flow at maximum engine rpm with a little higher back pressure.

On a street machine with relatively low valve overlap, positive differential will help scavenge the residual gases, but raw fuel will not be blown into the exhaust manifold.

Dynamometer Matching—The final matching of a turbine-housing A/R or nozzle area to obtain best overall performance is often done on a dynamometer. If the engine is to be used on a truck, sport-fishing boat or other steady-state application, dyno testing is undoubtedly the best way.

Racing Applications—Using a dynamometer for matching is fine for street-driven cars and trucks. However, most dynamometers are not suitable for matching a turbocharger to an engine to be used in a road or circle-track racecar. The exception is noted in the sidebar, page 43.

After the size of compressor has been established, the turbine housing should be matched to the track conditions just as final gear ratios are matched.

Turbine-Housing Changes—Most turbochargers available to kit builders or individuals have a family of turbine housings with various A/R sizes. This is not true in all cases; some models have only one or two turbine housings. It is a good idea to make sure several turbine-housing sizes are available for the turbo you plan to use.

Changing turbine housings on turbochargers takes about five minutes, including removing and replacing the inlet and outlet ducting. Insulated gloves are recommended. If V-band

Small diesel turbo was seen on a Buick-powered competition roadster in Flint, Michigan, in 1962. Two large 4-barrel carburetors are mounted on a "trunk" manifold to the compressor inlet. Diesel-type turbo had no seals to protect against gasoline flowing through compressor into turbo bearing, so turbo bearing failed. Junction box connected right and left exhaust manifolds next to turbo. This is a good example of mismatching—in turbo sizing, construction and carburetion requirements.

inlet and outlet connections are used as are quick disconnects for the oil system, a turbocharger can be replaced in less than a minute.

SUMMARY

Figure out the air requirements for your engine at the required speed. On a compressor map, this amount should fall between the surge line and the 60% efficiency line.

If boost pressure is too high, use a turbine housing with a larger A/R. Tuning the installation with the correct turbine-housing A/R is essential to make the turbocharger work correctly. Novice installers of turbo have nearly always blamed the unit for producing too much or too little boost when all they had to do to make it right was to change the turbine housing. Because the information has not been common knowledge in the past, this part of making the turbocharger installation work has been completely overlooked unless the tuner was able to get help from someone with turbo experience.

For this reason, it is important to buy your turbo from a distributor or manufacturer with various turbine housings in stock. They should also agree to work with you by swapping turbine housings until the installation is perfect.

IF YOU DON'T LIKE THE MATH

A good percentage of potential turbocharger users will read the details I've described and say, "But how do I know which turbocharger to use for my application?"

The people who sell turbochargers for special applications are anxious to make money. So they have gone through the process for just about any engine or application you can imagine.

Paul Uitti, Jon Meyer, Andy Johnston and Gale Banks have all spent years matching turbochargers to engines. They have contributed information for this book on matching turbochargers to various engines and applications. This is shown in Tables 3, 4, 5, 6 and 7.

No doubt, someone will ask why such-and-such a turbocharger is not on the list. Several manufacturers make many models of turbochargers, but only certain ones are readily available in frame sizes with families of interchangeable turbine housings. So, if you have a model not in these tables and it needs an overhaul, you might have trouble finding parts. Those shown in the tables are available *and* repairable.

If you cannot find your engine listed in the tables, one with the same displacement and valve-train type will normally take the same size turbocharger.

TESTING WITH THE VEHICLE

For those of us who do not have the money to spend for a programmable dyno, there is a less-expensive way to match a turbocharger to an engine in a vehicle.

Renault Turbo and Rally Engine Comparison				
RPM	Rally Torque	Turbo Torque	Rally HP	Turbo HP
3600	98	210	67	144
4000	106	229	81	174
4400	103	234	86	196
4800	107	232	98	212
5200	109	228	108	226
5600	108	224	115	239
6000	102	223	116	255
6400	95	214	116	261
6800	88	201	114	260
7200	76	192	104	263

Horsepower and torque figures for turbocharged IMSA Renault R5 prepared by Katech, Inc. compared to naturally aspirated rally engine. Both are production-based 1400cc pushrod fours. Basically a turbocharged rally engine, IMSA engine has more than double the horsepower and torque over entire rpm range.

TABLE 3

Suggested Turbochargers for Low Boost (Max. 7 PSI) V8 Engine Applications ns
Use 1 or 2 Turbos with boost control

DISPL IN³	RPM	CFM @ 1.5 PRESSURE RATIO	WARNER - ISHI MODEL	TRIM	RAJAY MODEL	ROTO-MASTER TO4B TRIM	TURBINE A/R
283	5000	429	2-RHB5	393B	2-301F70	2-(Y-4)	N-.40
					or	or	
					1-301E90	1-(V-2)	P-.81
305	5000	462	2-RHB5	393B	2-301F80	2-(Y-4)	N-.58
					or	or	
					1-301E10	1-(V-2)	P-.81
327	5000	494	—	—	2-301F90	2-(Y-4)	N-.69
						or	
						1-(V-2)	P-.81
350	4500	455	2-RHB5	393B	2-301F80	2-(Y-4)	N-.58
					or	or	
					1-301E10	1-(V-2)	P-.81
400	4000	481	2-RHB5	393B	2-301F90	2-(Y-4)	N-.69
						or	
						1-(V-2)	P-.96
427	4000	507	—	—	2-301F90	2-(Y-4)	N-.81
						or	
						1-(R-11)	P-.96
454	4000	533	—	—	2-301F90	2-(S-4)	O-.69
						or	
						1-(R-11)	P-1.15
500	4000	585	—	—	2-301E90	2-(S-4)	O-.81
						or	
						1-(R-11)	P-1.35

TABLE 4
Suggested Turbochargers For Moderate Output (Max. 15 PSI)
V-8 Engine Applications
Use 2 Turbos with boost control

DISPL IN^3	CFM @ 2.0 RPM	PRESSURE RATIO	RAJAY MODEL	ROTO-MASTER TRIM	TO4B TURBINE A/R
283	7000	713	2-301F80	2-(S-4)	(O-.69)
305	7000	760	2-301F90	2-(S-4)	(O-.81)
327	7000	806	2-301E70	2-(S-4)	(O-.81)
350	6500	806	2-301E70	2-(S-4)	(O-.81)
400	6500	915	2-301E80	2-(S-4)	(O-.96)
427	6500	977	2-301E90	2-(S-4)	(O-.96)
454	6500	1023	2-301E10	2-(V-2)	(P-.81)
500	6500	1163	2-301E10	2-(V-2)	(P-.96)

TABLE 5
Suggested Turbochargers for High Output (Max 25 PSI)
V8 Engine Applications
Use 2 Turbos with boost control

DISPL IN^3	CFM @ 2.7 RPM	PRESSURE RATIO	RAJAY MODEL	ROTO-MASTER TRIM	TO4B TURBINE A/R
283	7000	822	2-301F90	2-(S-4)	O-.81
305	7000	883	2-301F90	2-(S-4)	O-.81
327	7000	935	2-301E90	2-(S-4)	O-.96
350	6500	936	2-301E90	2-(S-4)	O-.96
400	6500	1078	2-301E10	2-(S-4)	P-.96
427	6500	1148	2-301E10	2-(V-2)	P-.96
454	6500	1212	2-301E13	2-(V-2)	P-.96
500	6500	1349	2-301E13	2-(V-2)	P-1.15

They say, "Necessity is the mothe of invention," so naturally somebod came up with a way to measure horse power in the vehicle. Depending o the amount of information desired this can be a simple or elaborat setup. Either way it is less costly tha a programmable dynamometer. I ca it the Griffin System because Dic Griffin was the first to tell me how t do it.

The basic idea is to use the weigh of the vehicle to resist th acceleration. Many enthusiasts don' like this method because it doesn' give as high a reading as an engin dynamometer. Horsepower measure by this method is *whee horsepower*—what's actually delivere to the road.

With a dynamometer, the metho often used is to rev the engine an catch the needle at the end of it swing. This gives great powe readings, but they're less tha reliable.

Even when reliable readings ar made at the rear wheels, there is tendency to add 30% or more fo "power-train losses." I have run a engine on an engine dynamomete and then put it in a car with a manua transmission. I found only about 4% difference in the output. Even with a automatic transmission, losses are no as great as some would like us t believe.

Passenger cars are rated fairl honestly these days. But a few year ago, the 210-HP car you bought onl put out about 117 at the rear wheel We liked to believe the higher horse power, so we charged the loss agains the drive train. It just wasn't so! Tha 210-HP engine never saw much mor than 150 HP at the flywheel—eve with the fan disconnected!

TABLE 6
Suggested Turbochargers For 4 and 6 Cylinder Engines:
One Turbo per engine with boost control. Maximum boost pressure 7 PSIG

ENGINE SIZE RANGE IN^3	MAXIMUM SPEED RPM	CFM @ 1.5 PRESSURE RATIO	WARNER - ISHI MODEL	TRIM	RAJAY MODEL	ROTO-MASTER TO4B TRIM	TURBINE A/R
60 - 90	7000	180	RHB5	311A	302B-25	Y-4	N-.30
90 - 130	6500	240	RHB5	393B	302B-40	Y-4	N-.40
130 - 170	6000	308	—	—	301F-70	S-4	O-.40
170 - 210	5500	360	—	—	301E-70	S-4	O-.58
210 - 250	5000	405	—	—	301E-80	S-4	O-.69
250 - 300	4500	435	—	—	301E-90	V-2	P-.58

TABLE 7
Suggested Turbochargers For 4 and 6 Cylinder Engines:
One or Two Turbos per engine with boost control. Maximum boost pressure 20 PSIG.

ENGINE SIZE RANGE IN³	MAXIMUM SPEED RPM	CFM @ 2.3 PRESSURE RATIO	WARNER - ISHI MODEL TRIM	RAJAY MODEL	ROTO-MASTER TO4B TRIM	TURBINE A/R
40 - 60	10,000	198	RHB5-311A	302B-25	Y-4	N-.3
60 - 90	10,000	297	RHB5-393A	302B-40	Y-4	N-.4
90 - 130	9,000	363	2-(RHB5-311A)	301F-60	Y-4	N-.58
130 - 170	8,000	446	2-(RHB5-393B) or 2(302B-40)	301E-70 or 2(301B-40)	S-4 or 2(S-4	O-.4 N-.3)
170 - 210	7,000	495	2-(RHB5-393B) or 1(301B-60)	301E-80 or 2(301B-60)	S-4 or 2(Y-4	O-.58 N-.4)
210 - 250	6,000	512	2-(RHB5-393B) or 2(301B-70)	301E-90 or 2(301B-70)	S-4 or 2(Y-4	O-.69 N-.4)
250 - 300	6,000	627	2-(RHB5-393B) or 2(301F-60)	301E-10 or 2(301F-60)	V-2 or 2(S-4	P-.58 O-.58)

SUPERFLOW PROGRAMMABLE DYNAMOMETER

Almost any dynamometer can be used for turbocharger/engine matching if the application is steady-state. For road or racing applications, it is essential to use a dynamometer that can simulate acceleration conditions. Until recently, only large engine manufacturers with extremely expensive dynamometer facilities could run such tests.

The SuperFlow microprocessor-controlled engine dynamometer can be programmed to provide acceleration rates from 25 to 2000 rpm/second. Its response time will handle the large jump in power output that occurs as the turbo begins to supply a significant amount of boost. Data is printed out at the end of the test or can be observed on the screen of a cathode-ray terminal.

At a cost of approximately $17,000, this dynamometer has become an essential part of the testing facility for a number of serious engine builders and tuners. Electramotive, Reeves Callaway, RC Engineering and HPBooks are all using these units for their development work. SuperFlow Corporation is at 3512 N. Tejon, Colorado Springs, Colorado 80907.

A few years ago we turbocharged a 4000-pound Oldsmobile with an automatic transmission. When it turned the quarter at over 103 mph on factory tires, charts showed this required almost 400 HP. We never measured more than 275 HP at the rear wheels.

The point is, don't be disappointed if rear-wheel horsepower is not as great as you expected. If you double the horsepower, the increase is 100% no matter how you measure it.

HOW TO MEASURE HORSEPOWER

1. Weigh the car and occupants accurately.
2. Install a tachometer.
3. Calibrate the speedometer.
4. Install an accelerometer.

The car and occupants may be weighed at any public scale. The tachometer can be calibrated by a strobe light or a revolution counter.

Calibrate the speedometer by driving at a constant speed on a highway equipped with mile posts. Use a stopwatch to measure the exact time for a given distance. The longer the distance, the more accurate the calibration. Actual speed is obtained by dividing the distance by the time.

Mount a reliable accelerometer in the vehicle. Don't use an aircraft-type; the divisions are too small. The maximum scale reading should not be greater than 1 G. Most readings will be below 0.3 G if the vehicle is a street machine.

After all the calibrations have been completed, the same two people included with the weight of the car will run the tests. Many runs must be made to obtain a power curve, but I will go through the entire process for only one run.

To find maximum horsepower at a specific rpm, first determine the road speed necessary to obtain that rpm in high gear. If it is above the legal limit, drop down a gear.

Starting at a speed well below that required, open the throttle all the way. Keep one eye on the speedometer. When it reaches the required speed, call out "mark" or some other predetermined word. The other occupant observes and records the accelerometer reading.

Continue accelerating to about 5 mph above the required speed. Put the transmission in neutral and start coasting. When the speed again reaches the required mph, again call out "mark." This time the other occupant observes and records the deceleration reading from the accelerometer.

The second reading is taken to measure power lost to air resistance and road load. Add the two readings to determine the total acceleration delivered by the tires to the road.

These readings will be accurate as long as there is no change in grade or wind direction during the interval between the readings. It is best to make several runs in each direction and average the readings.

The next step is to make use of the readings. You'll have to go back to a couple of formulas from high-school physics:

$$F = MA$$

or

Force = Mass x Acceleration

$$Mass = \frac{Weight}{Acceleration\ of\ Gravity}$$

Acceleration due to gravity is equal to 32.2 ft/sec².

If the accelerometer reading is in G units, which is:

$$\frac{\text{Acceleration of Vehicle in ft/sec}^2}{\text{Acceleration of Gravity in ft/sec}^2}$$

then

F = Weight x Accelerometer Reading

This force is the actual effort applied to the road by the tires. If the speedometer reading is converted to ft/sec, horsepower can be calculated directly.

Velocity (ft/sec) = Velocity (mph) x 1.467

One horsepower equals 550 lb-ft/sec, so road horsepower is:

$$HP = \frac{\text{Force x Velocity}}{550}$$

A small calculation shows how simple this is.

Weight of car and occupants	3650 lb
Corrected road speed	60 mph
Engine speed	2800 rpm
Full-throttle acceleration	0.12G
Neutral deceleration	0.06G
Total acceleration	0.18G

$$F = MA$$
$$= \frac{3650 \times 0.18 \times 32.2}{32.2}$$
$$F = 657 \text{ lb}$$

$$\text{Velocity} = 60 \times 1.467$$
$$V = 88 \text{ ft/sec}$$

$$HP = \frac{657 \times 88}{550}$$
$$HP = 105.1 @ 2800 \text{ rpm}$$

Engine torque can then be calculated.

$$HP = \frac{T \times rpm}{5250}$$

or

$$T = \frac{5250 \times HP}{rpm}$$
$$= \frac{5250 \times 105.1}{2800}$$

Torque = 197 lb-ft

If the road-load-horsepower readings are to be done regularly and extreme accuracy is desired, it's worthwhile to invest in some additional equipment.

In this case, 16mm movie camera can be mounted in the car to photograph the instruments during acceleration runs. Film is viewed on a small screen used for editing. Instruments measure compressor inlet and outlet pressure, time in 1/100 seconds, acceleration, engine speed and road speed.

The biggest single advantage of the camera over direct observation is

Jim Kinsler's instrument package for on-track testing: Video camera records instruments for track-side viewing and analysis. Digital and analog instruments supply an array of engine data.

objectivity. The camera is not biased—it will not slant readings according to the observer's preference. Anyone modifying an engine tends to read higher figures than actually occur.

Jim Kinsler has had good success using a video camera to record results on a portable videocassette recorder. Kinsler has done a lot of work on intercooling, and notes that it's extremely hard to dyno test a turbocharged, intercooled engine. These must be tested in the field or on the track because it's nearly impossible to judge (and adjust for) the exact air flow over the intercoolers in a dyno room.

Using the videocassette recorder also allows observing the results immediately after a test run.

If possible, readings of the naturally aspirated engine should first be taken and plotted. This will give a *base line* before the engine is turbocharged, Figure 4-12.

After the turbocharger is installed and initial settings for idle speed and ignition timing have been done, rerun the dynamometer tests. You can now

determine where there has been a power gain and how much.

If the power curve looks like Figure 4-13, it means the compressor is the right size but the turbine housing has been matched for an increase in power at the top end only. Using a turbine housing with a smaller A/R should give more boost at the low speeds end, Figure 4-14.

A too-small turbine housing may push the boost out of sight at high speeds. This is often fatal to engines, so small turbine housing should be avoided. A control or one of the restrictions mentioned earlier should be used.

When a turbocharger is properly matched, a horsepower curve similar to Figure 4-15 can be established. Please remember these curves are for *comparison only* and do not represent a specific engine. A 100% increase in road horsepower over a broad speed range is not only possible, it has been done many times.

This method of matching is fine for cars and trucks that are to be used on the street but is impractical for road or track racers.

5 Diesel

First turbocharged car to race at Indianapolis was Freddie Agabashian's Cummins Diesel. He put car on pole in 1952. Photo courtesy of Indianapolis Motor Speedway.

When the first edition of this book was published, the vast majority of turbocharged engines were diesel. Almost all of the installations were made by engine manufacturers—people well versed in the procedures required to obtain a good match.

Since that time, many light- and medium-duty vehicles have been manufactured with naturally aspirated diesel engines. Considerable interest has been shown in turbocharging them for better fuel economy, more power or both.

DIESEL ENGINES

In Chapter 1, I went through the four strokes of a four-stroke, spark-ignition engine. The same illustrations are used here for a diesel engine, but the similarity ends there.

Not shown in Figure 1-1 is the carburetor or fuel-injection system. Its operation is described in Chapter 6 and basically does two things. It mixes the air and fuel to produce a combustible charge to suit the conditions of the instant. It throttles the mixture to control engine power.

Ordinarily a diesel engine has no throttle. It receives a full charge of air regardless of the power output. Instead of having a spark plug to ignite the charge, the fuel is injected at high pressure at the top of the compression stroke.

Diesel engines have very high compression ratios, up to 22:1. The air at the end of the compression stroke is hot enough to ignite the diesel fuel as it enters the combustion chamber. Because it burns as fast as it is injected, the engine will not detonate at high output.

The power output of a diesel engine is a function of the fuel-injection system. Consequently, *turbocharging a diesel engine that was running clean naturally aspirated will have little or no influence on the power output.* If more power is desired, the pump must be recalibrated to deliver more fuel per stroke at the right time.

WHY TURBOCHARGE?

Turbocharging the spark-ignition engine offers improvements in size, weight and—sometimes—fuel econo-

Sanitary turbocharger installation in Toyota Land Cruiser is by TurboFrance. Output is increased by 30%.

Engine in my Datsun Z is a Roto-Master turbocharged 200-CID Nissan Diesel. Performance equals that of original 2800cc (171-CID) gasoline engine. Mileage achieved with overdrive is 30 mpg.

my when compared to naturally aspirated engines with the same power output. Turbocharging the diesel offers these improvements:

1. Smaller size
2. Lighter weight
3. Better fuel economy
4. More power
5. Altitude compensation
6. Reduction or elimination of smoke
7. Less noise
8. Lower emissions
9. Lower operating temperature
10. Automatic spark arresting.

These are not idle claims. There are good reasons for each benefit.

Forcing more air into an engine allows more fuel to be burned to increase power. This allows using a smaller engine for the same power output. A smaller engine also lowers overall weight.

Adding a turbocharger to a diesel does more than increase the amount of air available for combustion. The turbocharger also improves combustion efficiency by increasing turbulence in the combustion chamber.

Greater turbulence gives more complete combustion and more power for a given amount of fuel.

Additional air also allows more fuel to be injected to give more power.

Smoke Control & Power Rating—A naturally aspirated diesel is set up to operate with a small amount of excess air, or run slightly lean—usually around 18:1 air/fuel ratio—at *rated* power so it won't smoke excessively. Rated power is the manufacturer's quoted power output for an engine when it is *calibrated,* or tuned, to operate at a specific altitude and ambient temperature.

The mixture at maximum power will get richer as the air gets *thinner*—less dense. At about 1000 feet, the average naturally aspirated engine starts to run out of excess air, resulting in a richening of the fuel/air mixture. Unless the engine is *derated,* or recalibrated for less power at the higher altitude, the engine will smoke at maximum power.

A turbocharged diesel can compensate for less-dense air at higher altitudes in two ways. First, a turbocharged diesel usually has at least 50% excess air at sea-level-rated power. Second, gage pressure remains approximately constant with altitude.

This combination allows a turbocharged diesel to go to an altitude of at least 7000 feet before any derating is necessary. Some turbocharged diesel engines can operate above 11,000 feet without being derated.

A naturally aspirated engine is usually rated at the power where the exhaust smoke is just visible. Recent emission laws have caused many naturally aspirated diesel engines to be derated—fuel/air mixture leaned—even at sea level. Previously acceptable smoke levels are now considered excessive.

With the additional air available, a turbocharged diesel may never smoke. With a correctly matched turbocharger, a diesel engine will reach mechanical or thermal limits before smoking.

Controlling Noise—When turbochargers were first introduced on diesel engines installed in trucks and construction equipment, the muffler was often eliminated. It simply was not required to hold noise levels to an acceptable limit. But some states had laws requiring the use of a muffler regardless of the noise level.

To get around the problem, many turbocharger turbine housings had the words, **muffling device** cast into them. In other cases, a fake muffler was used with no internal baffling.

Noise laws, like smoke laws, have become more stringent in recent years. Consequently, mufflers are now usually used with turbocharged-diesel engines. The resulting sound reduction is considerably less than on a naturally aspirated diesel.

Hi-Tuned kit installed on Diesel Golf/Rabbit increased ouput from 50 to 75 HP.

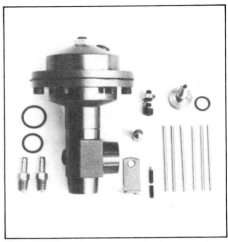

Spearco boost-pressure device, used on their Rabbit Diesel kits, prevents smoke due to overfueling when insufficient air is available.

Emission Controls—The addition of a turbocharger to a diesel does not seem to have much effect on NO_x (oxides of nitrogen) emissions. The additional air available makes significant reductions in both HC (hydrocarbons) and CO (carbon monoxide). In some cases the CO reading is so low it is below the accuracy of the measuring equipment.

Exhaust Temperature—A typical naturally aspirated diesel will reach the smoke limit and 1300F exhaust temperature at about the same power output. Turbocharging the same engine with a 50% increase in air flow and a 25% increase in power will lower exhaust temperature at the smoke limit to about 1100F to 1150F.

Additional air supplied by the turbocharger reduces engine operating temperature, even with the increased power. The reduction in operating temperature occurs only when the percentage increase in excess air is far greater than the percentage increase in power output.

Spark Arresters—All turbochargers used on engines up to about 1000 HP have a radial-inflow turbine. With this design, any sparks or glowing particles in the exhaust large enough to start a fire in a wooded or brush-covered area will not pass through the turbine wheel.

Centrifugal force from the turbine blades will drive larger particles back into the turbine housing. This process repeats until the original particle is reduced to particles small enough to pass by the turbine wheel with the exhaust gases.

TURBOCHARGER/DIESEL MATCHING

Even though it is important to have a proper match, matching a turbocharger to a diesel engine is less critical than matching one to a spark-ignition engine. The reason is that a diesel engine's output is controlled by the fuel-injection system.

Spark-ignition engine output is controlled by intake-manifold pressure—a function of the turbocharger. Because the spark-ignition engine must operate at or near a constant fuel/air ratio, doubling the intake-manifold pressure doubles the fuel flow. The combination roughly doubles power output.

With a diesel engine, doubling intake-manifold pressure will not increase fuel flow, so output remains about the same. Further output increases require recalibration of the injection system.

The one exception to this is if the naturally aspirated version smoked. In this case the turbocharger will clean up the smoke and give a slight power increase.

Injection Timing—Most people who have turbocharged spark-ignition engines quickly become aware of the importance of ignition timing for optimum results. Excessive spark advance causes power loss and detonation. Too much retard causes power loss, poor fuel economy and higher engine temperatures.

Injection timing on a diesel is just as important. Fuel injected too early causes combustion pressure to peak before TDC. This causes power loss

Known for their International Harvester diesel-tractor installations, Hypermax Engineering has a kit for turbocharging the 6.9-liter International Harvester/Ford diesel in Ford vans and pickups.

and possible engine damage. Fuel injected too late may result in fuel still burning at the bottom of the power stroke, even after the exhaust valve is opened.

In either case, exhaust temperature goes up and power goes down. This is true whether or not there is enough air to burn all of the fuel without smoke.

As calibrated by the engine manufacturer, an injection system is timed to accomplish two things. First, ignition should take place to produce peak cylinder pressure a few degrees ATDC. Second, all of the fuel should be burned long before the end of the power stroke.

To fulfill both requirements, the injection system must have some way to advance the injection timing as a function of engine speed. This is much the same as the centrifugal spark-advance

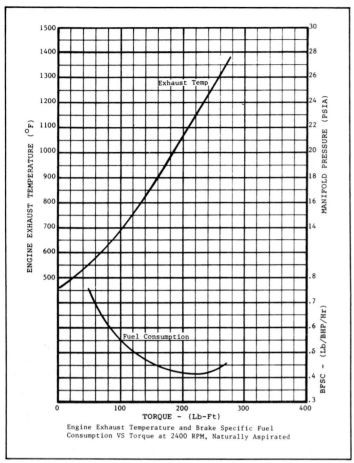

Figure 5-1—Part-load curve for naturally aspirated diesel engine

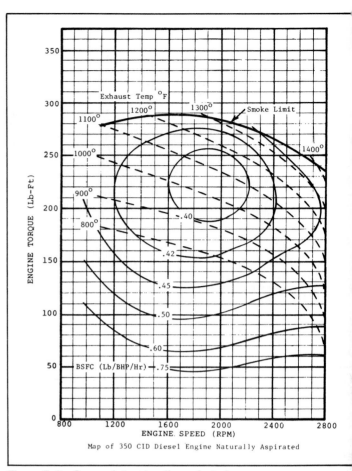

Figure 5-2—Cross plot, naturally aspirated

mechanism in a distributor or magneto on a spark-ignition engine. The fuel pump and injector must also be sized to inject the full charge in time for it to burn completely while it can still produce power.

If the injection system on a naturally aspirated engine has a very short injection time, it will probably be easy to increase the output at least 25% without running into trouble *as long as the pump has the capacity to deliver the additional fuel required.*

The manufacturers of some light-duty diesel engines for passenger cars have deliberately used injection pumps having very little excess capacity. In fact, if all the production tolerances are on the low side, pump capacity is barely enough to produce rated power.

The reason for this is to prevent overfueling of the naturally aspirated engine. Unfortunately, when turbocharging these engines, the low capacity makes it difficult to obtain the needed extra fuel with correct timing. Even if this is the case, turbo-

charging will eliminate smoking at high altitude.

Turbocharger Sizing—The procedure for matching a turbocharger for a specific application starts the same for a diesel engine as it does for a spark-ignition engine. But there is one difference: A diesel engine usually operates at a higher intake-manifold *pressure* and at a lower exhaust-gas *temperature* than the spark-ignition engine. As a result, turbine-nozzle size required for a diesel engine is usually smaller than that for a similar-size gasoline engine.

After the turbocharger has been selected, the method of matching on the dynamometer is different from that used with the spark-ignition engine. Power output is controlled by the fuel system, not intake-manifold pressure. A reasonable oversupply of air is not detrimental, and is usually beneficial with a diesel engine.

Establish Base Line—Turbocharger and engine matching is started by running the engine naturally aspirated on the dynamometer to establish a *base*

line. Several runs are also made to calibrate the fuel-injection for optimum engine performance before the turbocharger is installed.

Maximum torque at any speed is found by advancing the fuel-pump lever—to add more fuel—while the engine is loaded by the dyno until there is visible smoke from the exhaust. Injection timing is also varied to find maximum torque, although it has more affect on *brake specific fuel consumption* (bsfc). This is normally stated as pounds of fuel consumed per hour per brake horsepower.

To make a complete performance map of an engine, data is taken and part-load curves plotted at each speed. Data for a naturally aspirated diesel includes:

1. Engine speed
2. Torque
3. Exhaust-gas temperature
4. Specific fuel consumption
5. Smoke level
6. Injection timing (when applicable)
7. Ambient temperature and pressure

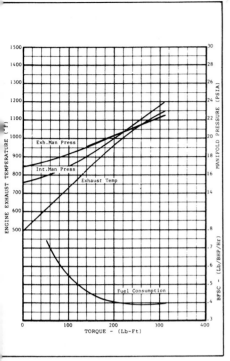

Figure 5-3—Part-load curve, 350-CID turbocharged-diesel engine

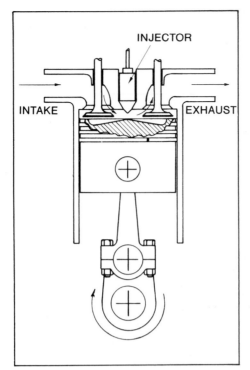

Figure 5-4—Valve-overlap period, direct-injection engine

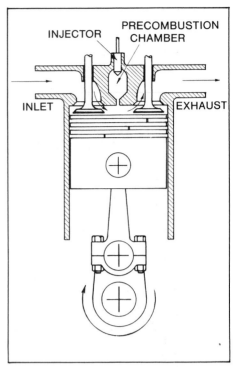

Figure 5-5—Valve-overlap period, precombustion-chamber engine

After maximum torque is established, part-load runs are made at 3/4, 1/2 and 1/4 of maximum torque and at idle. These readings establish the end of the exhaust-gas-temperature (egt) curve. The data are then plotted for each engine speed, Figure 5-1.

After all the runs, a complete engine-performance map is made by cross-plotting the data from the part-load curves. The cross-plot, Figure 5-2, shows the limits of operation of the engine. It also shows the ideal speed and load for best fuel economy.

If an engine is running on the borderline of smoke, it is often best to install the turbocharger with no change to the fuel system. Maximum output usually increases about 7% with a corresponding decrease in fuel consumption with no other change.

To verify the improvement, the engine is dynoed with the turbocharger installed. New data recorded at each speed is the same as for the normally aspirated engine, plus:

1. Intake-manifold temperature
2. Intake-manifold pressure
3. Exhaust-manifold pressure

Crossover—After the runs are completed, the part-load curve is again plotted, Figure 5-3. In most cases, there will be an exhaust-gas temperature above which the intake-

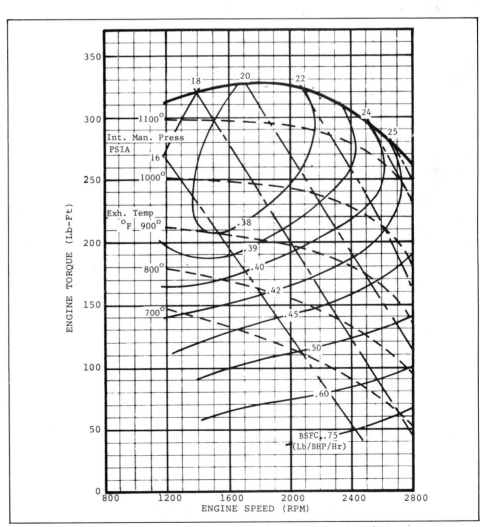

Figure 5-6—Cross-plot, turbocharged with naturally aspirated fuel setting

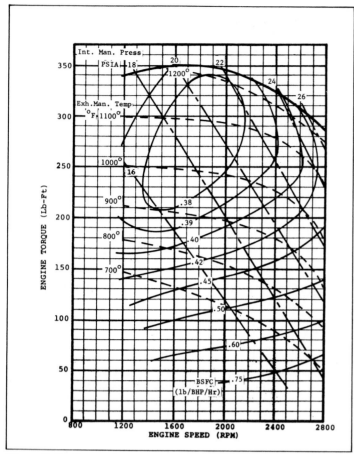

Figure 5-7—Cross plot, turbocharged with increased fuel setting

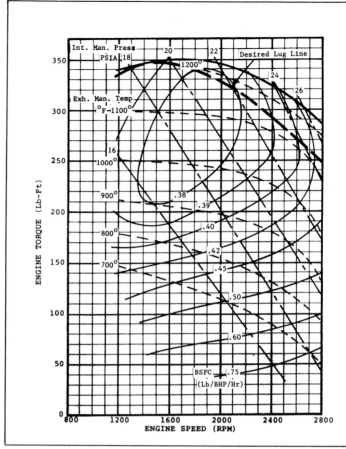

Figure 5-8—High-torque-rise lug line

manifold pressure will be higher than exhaust-manifold pressure. This temperature is often referred to as the *crossover temperature.*

When crossover occurs, it is possible to scavenge the combustion chamber to remove residual gases in the same manner as a naturally aspirated engine. Both valves are partially open during the overlap period, Figure 5-4, so air from the intake manifold flows into the combustion chamber and forces out residual exhaust gases.

The scavenging effect is not very great on a precombustion-chamber-type engine, Figure 5-5. The chamber is shrouded. Consequently, it is not possible to scavenge the precombustion part of the combustion chamber.

The cross-plot map of the turbocharged data, Figure 5-6, shows the increase in power and improvement in fuel consumption. For example, at 2400 rpm and 1000F exhaust temperature, the turbocharged diesel produces almost 230 lb-ft with a bsfc of 0.38. Under the same conditions,

Figure 5-2, the naturally aspirated model produced only 180 lb-ft with a bsfc of 0.42.

A turbocharger matched to give a maximum boost of about 8 psig at sea level should also allow the engine to be operated up to about 10,000 feet without derating.

The map, Figure 5-6, also shows the engine is no longer limited by smoke. And, exhaust-gas temperature is now considerably less than when naturally aspirated. This means additional power may be obtained by increasing fuel flow while still maintaining reasonable exhaust-gas temperatures and not generating objectionable smoke.

INCREASING POWER

An arbitrary limit for increasing power output without modifying the engine is 20 to 30%. This depends on the engine and how high it was rated naturally aspirated.

In any case, the rule of thumb is to keep the exhaust-gas temperature below 1300F. Higher exhaust-gas

temperatures will probably require piston cooling to prevent ring sticking.

Higher temperatures may even require lowering the compression ratio to keep the combustion-chamber pressure from going too high. Checking combustion-chamber pressure can only be done with elaborate equipment not available to the average shop.

Fuel-Injection Recalibration—The next step is to recalibrate the injection pump to obtain the higher horsepower. This should be done on a fuel-pump test stand by a specialist.

The pump must not only be timed correctly, it must be adjusted for the proper *torque backup.* Torque backup is the ratio of maximum torque compared to the torque at maximum speed. It is generally expressed as a percentage.

For example, in Figure 5-6, the torque at 2800 rpm is 262 lb-ft. Maximum torque is 328 lb-ft. The ratio of maximum torque to torque at maximum speed is 328/262 or 1.25. This is expressed as 25% torque backup.

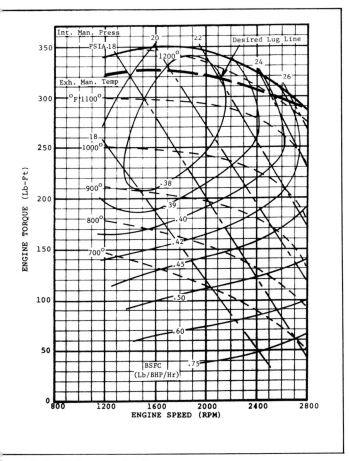

Figure 5-9—High-power lug line

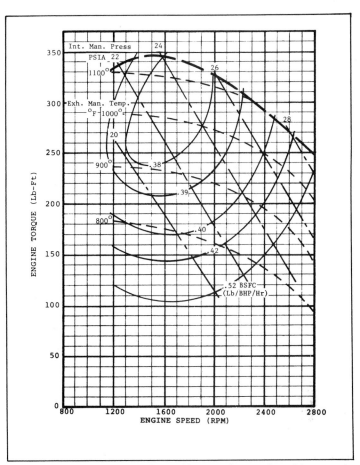

Figure 5-10—Effect of smaller turbine nozzle on exhaust temperature

Torque backup is important in non-stationary engines, which are usually required to run over a particular rpm range. If the load on an engine is increased the engine will slow down. But torque backup enables an engine running at maximum rpm to gain torque as it slows.

Rerunning Tests—Once the proper torque backup has been dialed in, the engine is rerun with the new pump setting. The data is again plotted and cross-plotted, Figure 5-7.

If there is an increase in exhaust-gas temperature without a corresponding increase in output, it is probably caused by retarded injection timing.

Too much advance will cause both high combustion pressure and a loss of power. For the aftermarket application, injection timing should not be advanced beyond the naturally aspirated setting because of the danger of exceeding safe combustion-chamber pressure.

If the map shows output and fuel consumption are acceptable, the engine and turbocharger combination can now be tested in its intended application. Any fuel setting that will keep the engine within its operating limits is acceptable.

Lug Lines—As previously noted, the fuel setting determines the power output. Each fuel setting within operating limits produces a particular torque curve, or *lug line*. The lug line is the actual torque curve that results from the calibration of the injection pump. The injection pump is calibrated to meet the specific application desired.

For example, a lug line shown on Figure 5-8 would be acceptable. The engine would operate at all times within the limits. The lug line on Figure 5-9 would also be considered acceptable.

For certain applications, acceptable is not good enough. If the performance map shows the system to be marginal at some point, Figure 5-8, the test should be re-run using the next smaller size turbine-nozzle or A/R, Figure 5-10.

If maximum power is required only at maximum speed, such as on a pump or electric generator, a larger nozzle area or A/R should be tested. Results are plotted in Figure 5-11.

Adjusting for Best Fuel Economy—On a truck or passenger-car application, it is best to have the engine speed, load and the best bsfc (brake specific fuel consumption) *island* all corresponding to the same road speed, Figure 5-12. On U.S. roads, that means 55 mph.

This is primarily done by gearing the vehicle correctly. But islands can be moved somewhat to the right or left on the chart by changing turbine-nozzle area.

The difference between a bsfc of 0.380 and 0.400 lb/HP-hr may not look like much, but it amounts to 5.3%. As an example I'll use a truck that gets 8 miles per gallon. If it operates in the best economy area only half the time and travels 100,000 miles per year, the savings could be $(0.053/2)(100,000/8) = 331$ gallons per year. This savings comes just from using the best turbine-housing match.

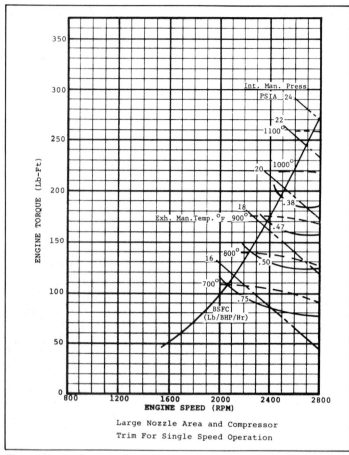

Figure 5-11—Larger nozzle area and compressor trim for single-point operation

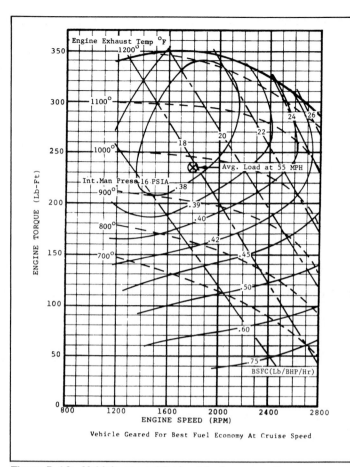

Figure 5-12—Vehicle geared for best fuel consumption

SPECIAL CONSIDERATIONS

In Chapter 2, I explained the difference between the compressor-end oil seals for diesel and draw-through carbureted engines. End seals must be considered with some diesels.

Some diesel engines use pneumatically operated injection systems. The Mercedes 240D used this type up to 1977 for U.S. cars—later than that in other parts of the world. The system contains a venturi and butterfly valve in the intake system. Vacuum signal from the venturi controls fuel flow.

If a turbocharger is added to this type of fuel system, the turbocharger must be mounted *between* the butterfly and engine. This mounting will require a mechanical seal behind the compressor impeller, much the same as with a carbureted engine.

If the compressor is mounted ahead, or upstream, of the butterfly, it will operate normally at part load. But at full load the venturi will not produce a vacuum signal. Without this signal the fuel pump will go to "wide open" and the engine can easily destroy itself from overspeeding.

This 10-cylinder MAN-type D2540 diesel is fitted with two KKK K28 turbochargers.

SUMMARY

Turbocharging a diesel engine to normalize for altitude or to increase the output by 25% is easier than turbocharging a spark-ignition engine. However, the finer points of matching should not be overlooked for best engine life and fuel economy.

6 Carburetion & Fuel Injection

Don Devendorf dominated the IMSA GTO class in 1983 in his Electromotive turbocharged Datsun Z. L28 six-cylinder turbocharged engine is intercooled and makes extensive use of electronics in its fuel-injection system. Photos courtesy of Electromotive Racing.

The purpose of a carburetor or fuel-injection system is to mix fuel with the air entering the engine. The fuel and air must be in the correct ratio for this mixture to burn efficiently in the combustion chamber. At the same time, the fuel/air mixture must not burn too hot because engine damage can result.

The task is the same for both naturally aspirated or turbocharged engines, with one exception. A turbocharged engine adds one more problem—that of the supercharged condition, where intake-manifold pressure is higher than ambient.

Fuel-System Basics—Modern carburetors differ in the way they work. But, in general, all have the following systems.

1. An enriching system to enable the engine to start and run when cold. This is the choke system.
2. An idle system to provide the correct fuel/air ratio to the engine when it is idling.
3. A main fuel system, which is in operation at cruise.
4. A power system to enrich the fuel/air mixture under extreme power or acceleration conditions.
5. An accelerator pump to give an extra shot of fuel to the mixture each time the throttle is opened. The additional fuel prevents a lag during the transition from the idle system to the main system, or the main system to the power system.

These carburetor systems or circuits—also duplicated in good fuel-injection systems—have been developed over the past 75 years. The modern carburetor is extremely reliable.

Progressive carburetors have a secondary barrel that is basically the same as the single barrel. Secondary barrels are added to reduce the pressure drop—increase air flow—at high engine speeds so the engine can reach maximum power. Progressive carburetors include the two-barrel type used on the 4-cylinder Chevette, Vega, Pinto, Mustang and Capri, and the four-barrel type used on many V8s.

Secondary barrels usually contain only as many systems as necessary for their operation. For instance, the secondaries ordinarily do not have a

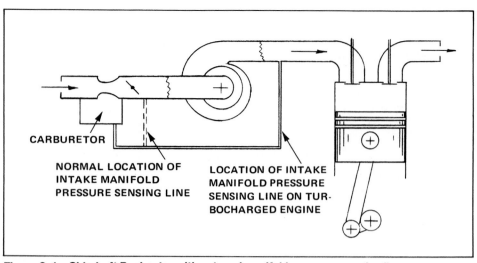

Figure 6-1—Sidedraft Rochester with external manifold-pressure sensing line

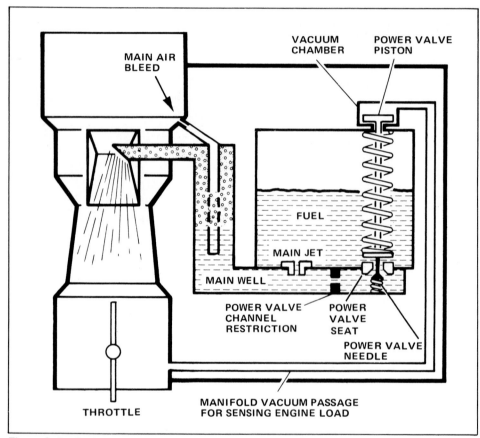

Figure 6-2—Carburetor power circuit showing power valve open

choke, and may not include a power system, accelerator pump or idle system.

ADAPTING THE CARBURETOR TO TURBOCHARGING

The main problems with carburetors and fuel-injection systems stem from one fact: They were invented before turbochargers became readily available.

Carburetors and fuel-injection systems were originally designed for naturally aspirated engines. They were not designed to supply a richer mixture when the intake-manifold pressure is above ambient, or 15 psia.

Looking back to the first turbocharged Corvair, Chevrolet used a Carter YH side-draft carburetor originally made for the six-cylinder Corvette. This carburetor had no way to sense intake-manifold pressure.

Two things were needed to keep the Corvair engine out of trouble when the boost went up to about 7 or 8 psig. One was to use a metering rod and jet that gave an overrich condition at full throttle. The other was a pressure-retard diaphragm on the distributor. This diaphragm retarded the spark about 10 crankshaft degrees when the intake-manifold pressure reached 2 psig.

The system worked well, but was a compromise. It never had the feel desired by many. Here was the problem:

An ordinary carburetor uses manifold vacuum to sense engine load. The power piston or diaphragm senses pressure immediately downstream of the throttle. When intake-manifold *vacuum* drops, the power piston supplies additional fuel.

Problems arise using this arrangement on a turbocharged engine, particularly if the turbocharger is mounted between the carburetor and engine. Partially closing the throttle when the engine is under boost will produce vacuum between the turbocharger and the carburetor. The increased vacuum will close the power valve while the engine is still supercharged. When this happens, the engine will run lean and probably detonate.

While Chevrolet was working on the turbo Corvair, Oldsmobile was developing its Jetfire, a turbocharged version of its 215-CID V8 F85 Cutlass engine. Oldsmobile's approach was quite different.

They started with a high-compression engine, but developed a carburetor specifically for a turbocharged application. This carburetor used two features well-suited to a turbocharged engine.

On the Oldsmobile, the power valve was not connected to the downstream side of the butterfly. It had an external line connected to the intake manifold downstream of the turbocharger, Figure 6-1. This prevented the power valve from being "confused" by vacuum between the carburetor and turbocharger.

The other modification Oldsmobile made was to add anti-detonant fluid to the carburetor when the engine was supercharged. The anti-detonant was necessary because of the engine's high compression ratio.

This same system will also work well with a low-compression engine. The anti-detonant can be added at a much higher manifold pressure. See Water Injection, Chapter 19.

Power-Valve Modifications—We have to live with carburetors designed for naturally aspirated engines, but several things can be done to adapt them to a turbocharged engine. Figure 6-2 shows the power circuit of a typical carburetor. Note that the sensing line senses vacuum just downstream of the throttle valve, as described previously.

This sensing line is normally a drilled internal passage. If the passage is plugged, a new line can be connected to the power diaphragm. This line can be routed to the intake manifold downstream of the turbocharger with a short length of steel or brass tubing.

On some carburetors, there is enough material in the wall to drill and tap for a tube fitting. On others, you'll have to drill a hole the size of the tubing OD and epoxy the tubing in place.

This modification only solves one problem—giving the correct signal to the carburetor power system. It does not do anything to enrich the fuel/air mixture when intake-manifold pressure is above atmospheric.

MIXTURE ENRICHMENT

There are several ways to provide the extra fuel without redesigning the carburetor. The device in Figure 6-3 can be added to the carburetor with the least effort. I have run this system on a turbocharged Corvair with good results.

Get the Hobbs pressure switch at an automotive parts-supply house. These come in various pressure ranges, but mine has a 0—4-psi range. The actual point—pressure—at which the switch closes can be adjusted with a small screwdriver.

The solenoid valve may be more difficult to find, however they should be available at recreational-vehicle outlets. These valves are normally used to switch from one fuel tank to another. Make sure the valve seat is made from a gasoline-resistant material.

The size of the orifice needed to add the extra fuel varies with the engine. I found 0.040 inch works pretty well for my uses.

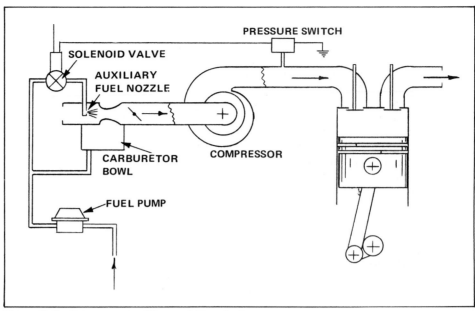

Figure 6-3—Pressure-sensitive enriching device

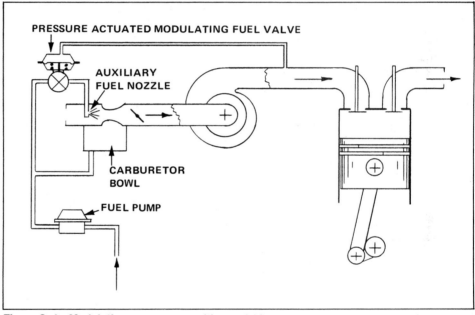

Figure 6-4—Modulating pressure-sensitive enriching system

This fuel-enriching device allows the carburetor power valve to be set at a fuel/air ratio that would normally be too lean for a supercharged engine. As soon as the intake-manifold pressure reaches about 2 psi, extra fuel is sprayed into the carburetor. This cools the engine and prevents detonation.

It is possible to drive all day without ever reaching the supercharged condition, so this modification also prevents wasting fuel. Unfortunately, it is an off-and-on system, so it is also a compromise. The engine will run too rich at 2- to 3-psi boost and too lean at 8 to 10 psi.

This system can be made to meter fuel, Figure 6-4. In this case, the amount of fuel added is controlled by a diaphragm that senses intake-manifold pressure. The fuel/air mixture is gradually enriched as intake-manifold pressure increases.

Although it can be made to do a better job than the pressure switch and solenoid, this system requires considerably more development. Reeves Callaway markets this type of enrichment device.

Callaway "Turbofueler" provides progressive fuel enrichment as boost increases. Fuel-enriching device works on same principle as that shown at right. Photo by C.E. Green.

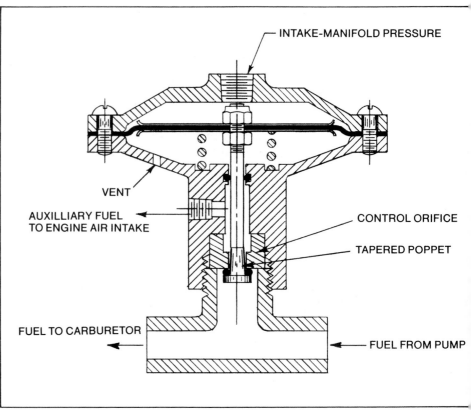

Figure 6-5—Modulating fuel-enriching valve

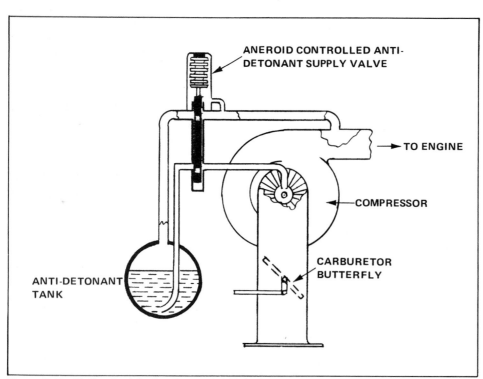

Figure 6-6—M. Goudard device, Patent 2,290,610

Just to make sure we don't think we are "reinventing the wheel," Figure 6-6 shows a device patented by Maurice Goudard in 1942, Patent 2,290,610. Although this device was designed for an aircraft engine, it is not limited to one.

Even before that, two Englishmen, Fedden and Anderson, developed a method of introducing anti-detonant to a supercharged engine. A schematic drawing of British Patent 458,611 issued in 1936 is shown in Figure 6-7. These are shown for historical reasons, but point out that supercharging had the same thermodynamic problems decades ago as it does today.

Progressive Carburetors—If a progressive two- or four-barrel carburetor is used, fuel enrichment can be approached in a completely different manner. The power valve should still be modified, but fuel enrichment can be done completely in the secondary.

Progressive-type carburetors have all the systems necessary to run the engine in the primary barrel or barrels. As mentioned before, the secondary is used to reduce the pressure drop through the carburetor at high engine speeds.

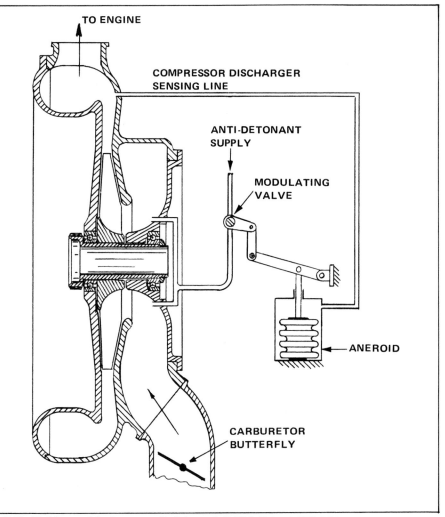

TO ENGINE

COMPRESSOR DISCHARGER SENSING LINE

ANTI-DETONANT SUPPLY

MODULATING VALVE

ANEROID

CARBURETOR BUTTERFLY

Figure 6-7—Fedden and Anderson anti-detonant, British Patent 458,611

Arkay made this installation on the Mazda 626 pace car for the Champion Spark Plug racing series.

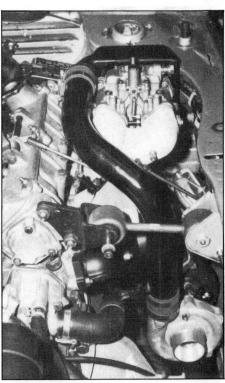

Janspeed managed to install turbocharger, blow-through Weber 40 DCOE carb and waste gate in small Fiat X19 engine compartment. Webers work well on turbocharged engines, but parts and tuning can be a problem.

Pressure drop at the carburetor represents a restriction in the intake. A 2-psi drop through a carburetor means the engine has 2-psi less atmospheric pressure to fill the cylinders.

On a naturally aspirated engine, if ambient pressure is 30 in.Hg, a 4-in.Hg drop means the engine can only put out (30-4)/30, or 87% of its potential.

When the secondaries of a progressive carburetor open, this restriction drops to practically nothing. The engine can then achieve its full potential as far as breathing capacity is concerned.

The problem with carburetor restriction does not exist on a turbocharged engine. Within reason, any pressure drop through the carburetor can be made up by the turbocharger. It is possible to have an 8-in.Hg (4-psi) drop through the carburetor and still have a manifold pressure of 10 or 15 psig.

The point I am making is this: A progressive carburetor is not necessary on a turbocharged engine to reduce the pressure drop and increase power. But its two separate fuel/air systems can be used to correct the fuel/air ratio when the engine is supercharged.

The secondary system on a progressive carburetor can be actuated by several different methods. A simple mechanical linkage can start to open the secondary when the primary is three-quarters open. Or, the secondaries may be diaphragm-operated using the vacuum drop through the primary venturi as a vacuum source.

Diaphragm-operated carburetors usually have an interlock to help pull the secondary throttles closed as the primaries close. This type of actuation is used on the Holley 4150/60-series

Turbocharging a fuel-injected engine usually creates the fewest plumbing problems. Six-cylinder BMW uses Bosch fuel injection with air-flow sensor that compensates for additional air supplied by turbocharger. Neat installation by Callaway. Photo by C. E. Greene.

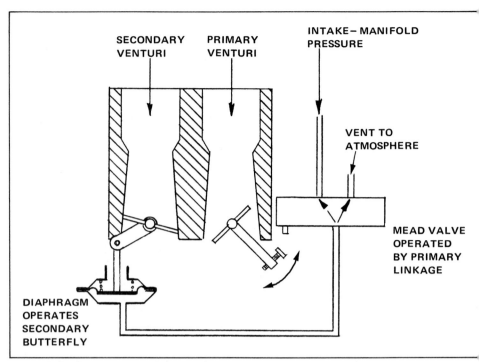

Figure 6-8—Progressive carburetor with pressure-operated secondary

Author's twin-turbocharged Corvair engine before it was installed. Diaphragm actuator for secondary throttles was rearranged to open secondaries with manifold pressure rather than primary vacuum.

carburetors and is easily modified for turbocharger use, Figure 6-8.

The interlock between the primaries and secondaries is removed completely. The diaphragm assembly is taken apart and the spring moved to the opposite side of the diaphragm. The linkage is reworked so pressure on the diaphragm opens the secondary throttle and the spring closes it.

With this modification, the secondaries open only when the engine is supercharged. Consequently, it is possible to select main jets and power valves for the primary to give the ideal fuel/air ratio for naturally aspirated conditions. Secondary-system jets can be sized for a rich condition more compatible with high intake-manifold pressures, or boost.

When a carburetor is modified this way, there is no mechanical linkage between the primary and the secondary butterflies. The engine operates using only the primaries until the intake-manifold pressure overcomes the spring in the diaphragm. This opens the secondary butterflies.

If no other modifications were made, the secondaries would not close when the driver's foot was removed from the accelerator pedal; the engine would continue to run supercharged on the secondaries alone. To prevent this, a small three-way *Mead valve* is placed in the sensing line and mounted on the carburetor.

The Mead valve is actuated only when the primaries are fully opened. As soon as the primaries close, the valve vents the secondary diaphragm to the atmosphere; the secondaries close immediately. This Mead valve is the same size and shape as a standard microswitch and is actuated in the same manner.

This setup gives a different feeling to the car than the standard setup. The secondaries will not open until manifold pressure reaches about 1 psig. When this happens, there is not only a sudden jump in manifold pressure but a definite surge in power.

BLOW-THROUGH OR DRAW-THROUGH?

Ever since turbochargers have been available for carbureted engines, the question whether to draw or blow through the carburetor has come up. There are advantages and disadvantages either way. Before knocking one or the other, remember excellent results have been obtained with both methods.

Table 8, this page, shows the advantages and disadvantages of blowing and drawing through the carburetor. Neither method is perfect, but either will do a good job when done correctly.

The biggest advantage of blowing through the carburetor is that it requires the least modification to the engine's intake system. Fuel lines, choke mechanism and accelerator linkage may all be left as is.

Turbocharger Capacity—Some say that blowing through the carburetor makes the turbocharger more efficient—*not so*. Turbocharger efficiency is a function of design, not carburetor location. If the peak adiabatic efficiency of a compressor is 75%, it will be 75% whether the carburetor is upstream or down.

When people say this, I think they actually mean capacity rather than efficiency. Due to the pressure drop through a carburetor, a turbocharger compressor downstream of the carburetor sees air below atmospheric pressure.

Because the centrifugal compressor is a cfm device, the maximum-flow potential in cubic feet per minute will be the same. But the maximum pounds per minute will be greater if the intake conditions never go below ambient.

Saying this in a different way, a slightly higher-capacity compressor is required to achieve the same maximum horsepower when drawing through the carburetor. If a large enough carburetor is used, this turbocharger-size difference is insignificant.

Compressor Seals—One big advantage of blowing through the carburetor is that it's not necessary to have a positive seal on the compressor end of the impeller shaft.

If a turbocharger is scrounged from a diesel engine, chances are it will have a piston-ring seal on the compressor end. This type of seal is not recommended on a draw-through car-

Another tight-fit problem solved: Corvettes are hard to turbocharge because of space limitations. Larson Engineering offers blow-through system with one or two AiResearch turbos, Blake waste gate, water injection and a special pressure-relief valve in the cast-aluminum adapter over the carburetor. Stainless-steel exhaust system is part of kit.

TABLE 8
Advantages and disadvantages of blowing or drawing through the carburetor.

Feature	Blow Through	Draw Through
1. Location of carburetor and linkage	No change	Must be moved
2. Fuel pump	Requires either extra pump or one compensated for compressor discharge pressure	No change
3. Positive crankcase ventilation	Must be moved to compressor inlet	No change
4. Fuel evaporation system	Must be equipped with check valve to prevent supercharging fuel tank	No change
5. Carburetor leakage	All holes and shafts must be sealed or complete carburetor boxed	No change
6. Distance from carburetor to cylinder	No change	Much longer
7. Carburetor float	Might collapse unless made of closed-cell plastic	No change
8. Turbocharger oil seal	Nothing special required	Must have positive seal on compressor end
9. Compressor surge	Can be a problem on deceleration	Not ordinarily a problem
10. Compressor size	Inlet pressure always atmospheric so maximum capacity always available	Inlet pressure below atmospheric so slightly larger physical size is required in some cases
11. Carburetor inlet temperature	Low temperature at low load, high temperature at high load	Constant temperature, regardless of load
12. Vacuum source for brakes, air-conditioning, etc.	Sporadic, requires check valve	Must be picked up between carburetor and compressor inlet

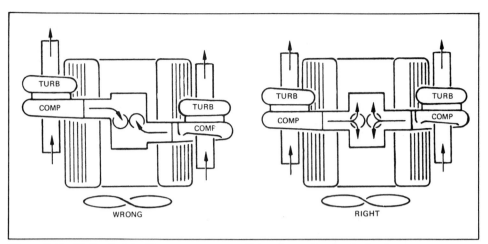

Figure 6-9—Method of ducting compressor outlets when using two turbochargers.

Duke Hallock, former high-performance coordinator and test-lab supervisor for AiResearch, has driven the same '37 Ford pickup since it was new. Ford 292 with AiResearch T-7 turbo is one of many engines he has installed in the chassis over the years. A plate perforated with 1/4-inch holes is used at original manifold flange to create turbulent flow for good mixture distribution.

buretor installation. It will not seal oil much beyond 5-in.Hg vacuum; a draw-through installation may produce as much as 29-in.Hg vacuum.

An engine with this type of turbocharger and installation will have a lot of blue smoke coming out of the exhaust. In addition, it won't take long to use up all the oil in the crankcase.

Don Hubbard suggests adding a second butterfly downstream of the compressor, as shown in Chapter 2, Figure 2-26. This was done on Indianapolis-type race cars using Schwitzer turbochargers. The second butterfly is linked to the main butterfly so the compressor seal never sees high vacuum.

Fuel Distribution & Intake Manifolds—The people who design engines do a pretty good job of designing intake manifolds. One of their major objectives is to provide good fuel distribution—the fuel/air mixture from the carburetor delivered to each cylinder in the same ratio.

When the intake system is left intact, fuel distribution remains about the same. But, when a turbocharger is placed between the carburetor and the manifold, fuel distribution will likely change.

The connections from the compressor to the intake manifold become critical. The fuel/air mixture comes out of the compressor in a swirl, which must be broken up before it enters the intake manifold. Turbulent flow is usually restored by placing a square-cornered plenum between the compressor-discharge pipe and the intake manifold.

If two turbochargers are used, as on a V8, the square plenum chamber is also important. Even more important, the compressor discharges should enter the plenum head-to-head rather than tangentially.

Looking at Figure 6-9, the left sketch shows two compressors mounted with their discharges entering the plenum tangentially. With this arrangement some cylinders will run lean while others run rich. Distribution can become so poor that leanness in some cylinders burns holes in pistons while richness fouls spark plugs in others.

In one case, when this happened the inlets were changed to the configuration shown in the right sketch. No further distribution problems were encountered.

Similar problems have occurred when only one turbo was used. Duke Hallock's 292-CID Ford installation, shown at left, directs the turbo outlet into a *doghouse* atop the intake-manifold carb flange. A plate with 1/4-inch holes had to be added to restore turbulent flow for equal mixture distribution.

An intercooler or a slightly longer tube between the turbo and the manifold might have accomplished the same thing. Any device between the turbo and the engine that will break up swirls or *laminar*—layered—flow into turbulent flow is definitely recommended.

Cast manifolds usually have enough sharp bends to promote turbulence. Fabricated intake manifolds must be carefully designed to prevent fuel separation.

Using individual inlet runners of specific lengths to get a ram effect at a certain engine rpm is not recommended. The turbocharger compressor should provide all the air needed. Chances are a tuned system will work well at one speed only and be detrimental at other speeds.

Sidedraft or Downdraft Carburetor?—Most turbochargers are designed to be run with the shaft horizontal. Consequently, a sidedraft carburetor is extremely easy to mount. There are, however, a few drawbacks.

For every sidedraft built there must be a million downdrafts. By sheer numbers, the average downdraft carburetor is further developed, easier to buy and service. There are exceptionally good sidedraft carburetors, but they are usually quite expensive.

"One drawback with most U.S. passenger-car downdraft carbs is that they are subject to mixture changes due to G forces in one direction or another," says Ted Trevor. He developed kits to adapt the Weber 40 and 45 DCOE sidedraft carbs to Rajay turbos.

Trevor claims only a little effort is required to get the Webers to provide correct mixtures under G loads encountered in slaloms, road racing, hill climbs—or even the fast maneuvering required for evasive action in everyday driving. As you will note from some of the accompanying photos, several installers have made good use of the Weber- and SU-type sidedraft carburetors.

Carburetor Heating—Another advantage of a blow-through installation is that all passenger cars have some way to warm the carburetor to prevent icing. On an in-line engine, carburetor heat is usually supplied by bolting the exhaust manifold directly to the intake manifold under the carburetor.

Some in-line engines use a water-heated spacer between the carburetor and the intake manifold. On some engines the entire intake manifold is heated by jacket water.

On a V-type engine, two exhaust ports (one in each head) are often connected by a heat-riser passage through the intake manifold. This allows hot exhaust gases to heat the intake manifold near the carburetor. Some V engines use jacket-water passages in the intake manifold or a

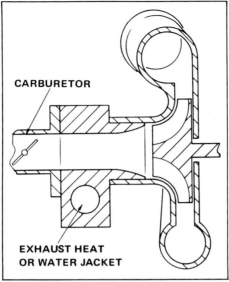

Figure 6-10—Sidedraft-carburetor spacer with built-in heater

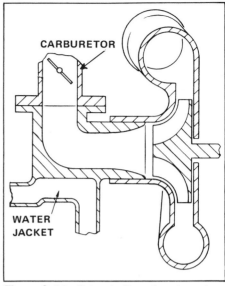

Figure 6-11—Downdraft-carburetor adapter with built-in heater

combination of jacket water and a heat riser.

When the turbocharger compressor is placed between the carburetor and the intake manifold, the carburetor is isolated from manifold heat. The carburetor can ice up or fuel can condense in the compressor housing.

Some turbochargers have been designed with water-jacketed compressor housings. Most are designed to keep the compressor housing as cool as possible to increase the volumetric efficiency.

To prevent carburetor icing, a spacer should be placed between the carburetor and the compressor, Figure 6-10. A drilled or cored passage through the spacer should be connected to the heater to allow hot water to flow through the passage. Carburetor-discharge temperature above 60F will normally prevent icing and condensation.

With a downdraft carburetor, a water passage should be drilled or cored on the bottom of the elbow, Figure 6-11. If fuel collects anywhere in the induction system, raw fuel will be carried into the cylinder when the throttle is opened. This can cause anything from stalling to overspeeding.

Fuel-Pump Modifications—One of the problems with blowing through the carburetor is maintaining fuel pressure. When the carburetor is pressurized, the fuel bowl is also under pressure. Air pressure in the fuel bowl acts against fuel-pump pressure.

The average fuel pump will not pro-

duce more than about 5-psig fuel pressure to the carburetor. If the turbocharger is delivering 10 psi to the carburetor, it is impossible for fuel to flow into the fuel bowl.

One way to overcome this problem is to add a high-pressure electric pump in series with the engine-driven pump. The electric pump must be located between the mechanical pump and the carburetor. Otherwise, the mechanical pump will act as a pressure regulator.

This arrangement will deliver enough fuel pressure when the engine is supercharged, but will also deliver high pressure at part load. This can cause severe fuel-control problems.

The additional fuel pressure may overcome the buoyancy of the inlet-valve float, causing fuel to pour through the carburetor and flood the engine. If the fuel goes through the engine and ignites in the exhaust, the exhaust pipe will act like a blow torch.

Overpressure can be prevented by operating the electric fuel pump with a pressure switch. The switch should start the pump only when the manifold pressure goes above atmospheric. This will work, but it costs money. It also adds another potential source of trouble and does not control fuel pressure smoothly.

The high-performance Holley electric fuel pump has a regulator mounted near the carburetor. One side of this remote regulator is vented to the atmosphere as a reference. A setscrew on the regulator is factory set to provide 6 psi to the carburetor.

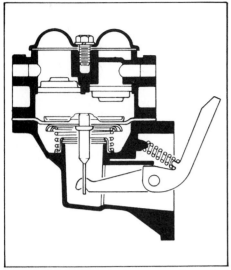

Figure 6-12—Typical automotive fuel pump

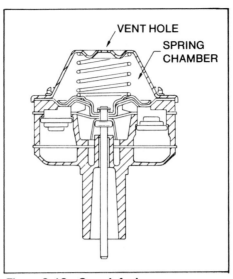

Figure 6-13—Corvair fuel pump

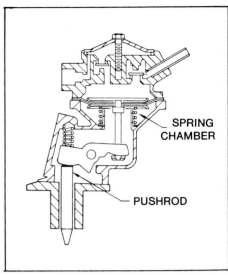

Figure 6-14—Cross section of VW fuel pump

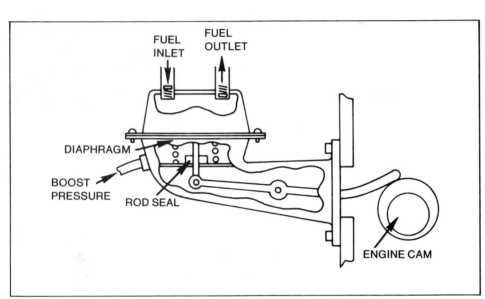

Figure 6-15—Diaphragm-type pump with rod seal

The pump itself is set to provide 14 psi to the regulator. The regulator vent can be hose-connected to the duct that feeds air to the carburetor. In the normally aspirated condition, fuel pressure will remain at 6 psi. Boost pressure will then automatically raise regulator-outlet pressure by the amount of boost.

Because the Holley pump produces 14-psi pressure, the maximum boost that can be accommodated with this pump/regulator modification is 8 psi. At any boost above 8 psi, the effective fuel pressure will be reduced by the amount that the boost exceeds 8 psi. For example, 10-psi boost will lower the fuel pressure to 4 psi—2 psi less.

Holley pumps and regulators can be made to exert more than 8 psi by increasing the spring force on the pump's bypass valve. Holley notes that this type of modification will shorten pump life, but it can be done.

Look at the cross section of a typical mechanical fuel pump in Figure 6-12. The rocker arm is operated by the engine cam and compresses a spring. When the rocker is on the heel of the cam, the spring provides the force to operate the diaphragm and pump the fuel. Pressure can be increased by installing a stiffer spring; but, like the electric pump, it will deliver higher pressure even when not needed.

Fuel pressure will remain constant as long as the flow is less than pump capacity. As fuel flows above pump capacity, fuel pressure becomes erratic.

Figures 6-13 and 6-14 are cross sections of Corvair and Volkswagen fuel pumps. The Corvair pump has its operating spring on the top side of the diaphragm. The operating spring on the Volkswagen pump is below the diaphragm.

With the Corvair pump, a fitting can be attached to the vent hole of the pump. A hose from the vent hole to the compressor discharge will apply boost pressure to the pump diaphragm. When the engine is supercharged, the spring will be assisted by compressor-discharge pressure.

With this modification, the Corvair pump will always deliver standard pump pressure *plus* supercharge pressure. The pump in effect becomes a pressure regulator.

The Volkswagen fuel pump has the spring on the bottom of the diaphragm. The rocker arm is driven by a pushrod, which is in turn driven by the engine cam. The pushrod fits loosely in a plastic guide.

If the pushrod is replaced with one that fits snugly in the guide—to prevent pressurization of the crankcase—the lower chamber of the pump may be pressurized. Again, a hose from the compressor discharge to the lower chamber is all that's needed. Then the pump will always deliver adequate fuel pressure.

The rod can occasionally stick in the up position, causing the pump to cease operating. When running

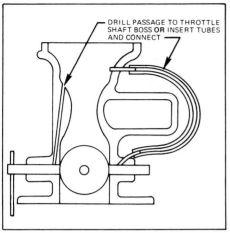

Figure 6-16—Air-sealed throttle shaft

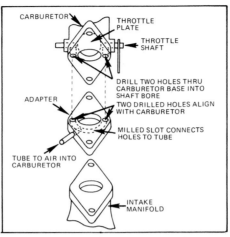

Figure 6-17—Adapter for carburetor-flange air seal

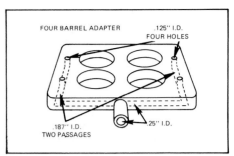

Figure 6-18—Adapter for 4-barrel-carburetor flange

Indy Car version of 90° Chevrolet V6 is based on small-block V8. Engine has aluminum block and angle-valve heads. Turbo blows through Hilborn fuel injection. Photo by Tom Monroe.

supercharged, however, pressure will tend to prevent sticking if the rod is very smooth.

This method may be used on any fuel pump in which the spring chamber can be sealed from the crankcase, Figure 6-15. It will not work where the spring chamber is open to the crankcase. Applying boost pressure to an unsealed spring chamber will pressurize the crankcase, which is neither practical nor safe.

Sealing Carburetor for Blow-Through Setups—After all the fuel-delivery problems have been solved, it may still be impossible to pressurize a carburetor. Some carburetor designs do not lend themselves to being sealed. In these cases, you will have to build a box around the carburetor and run the linkage in through a seal.

Don Hubbard has done quite a few blow-through applications and makes the following suggestions:

If the carburetor will be pressurized from the air-cleaner flange, any vents, including the float bowl, should vent to the air horn. This prevents air or fuel/air mixture from blowing outside the carburetor.

Floats should be plastic foam or Nitrophyl. Brass floats with flat sides may collapse under pressure or backfiring. They will then sink and flood the engine. Cylindrical brass floats, such as those used in SU carburetors, can usually withstand boost pressure.

To prevent collapse, a brass float can have a small hole drilled in it and plastic foaming liquid poured in. After the plastic foam has hardened, the hole can be resealed. Most carburetors come with foam floats or have foam replacements available. These will not collapse under pressure.

Don also suggests sealing the throttle shafts. The choke shaft only leaks dry air, so it requires no seal. However, the fuel/air mixture will leak out past the throttle shaft if it's not sealed.

Mechanical seals are not the best way to seal a throttle shaft; these cause excess friction. If the throttle has no linkage on one end, the shaft can be shortened and a plug pressed into the end of the bore. Otherwise, the best way is to use an *air seal* at both ends of the shaft.

Dry air from above the venturi can be ducted into the throttle-shaft bore. Air pressure above the venturi is slightly higher than below, so dry air leaks into the throttle bore. This blocks the fuel/air mixture, preventing it from escaping past the throttle shaft. This solves the problem without creating extra friction.

Some carburetor castings can be drilled for the modification. Some may require drilling and pressing in a small tube connected with flexible tubing, Figure 6-16.

A similar method Don uses to seal throttle shafts is with a special adapter flange, Figures 6-17 and 6-18. Holes drilled through the bottom of the carburetor flange carry air into the throttle-shaft bores. In a two-shaft carburetor—four-barrel or staged two-barrel—be sure the holes are the same size so one hole will not rob pressure from the others.

FUEL INJECTION & TURBOCHARGING

Turbocharging a fuel-injected engine has many of the same problems as blowing through a carburetor. This is because the injection nozzles are always on the pressure side of the compressor.

Some types of mechanical fuel injection have *aspirated nozzles*. Aspirated nozzles are vented to ambient air pressure to prevent manifold vacuum from drawing fuel from the nozzle at idle or under low-load conditions. This type fuel injection was used on early Chevrolet Corvettes and on some aftermarket injection systems.

When engines with this type of fuel injection are supercharged, pressurized air at the nozzle can prevent fuel from entering the port. To eliminate this problem, the aspiration vents

Throttle plate is immediately downstream of compressor discharge on this Indy Car (arrow). Arm at end of throttle-plate shaft controls fuel-metering block. Injector nozzles are at individual intake ports. Photo by Tom Monroe.

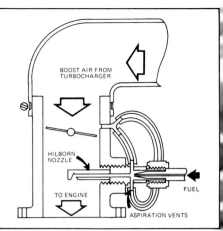

Figure 6-19—Fuel-injection aspiration venting

Turbo Carrera flat-six with KKK turbo pumps pressurized air to fuel-injection system. Waste gate allows quick turbo response. The 1975-77 model did not have an intercooler. Intercooler is standard equipment on '78 and later models.

must be connected to the compressor discharge, Figure 6-19. In addition, fuel-injection pressure must be increased to ensure good fuel flow at supercharged conditions.

As with carburetors, most designers of fuel-injection systems did not have turbocharging in mind. Therefore, they made no provision for adding extra fuel when intake-manifold pressure goes above ambient. Kit manufacturers use different methods to enrich the mixture.

When Rick Mack turbocharged the fuel-injected Porsche 914, he increased fuel pressure to ensure adequate flow at high boost. In addition, he added a manifold-pressure-modulated restrictor valve in the excess-fuel-return line from the injectors.

As intake-manifold pressure went above ambient, the valve would progressively restrict the excess fuel returning to the tank. This automatically increased fuel pressure at the injector nozzles and enriched the mixture. The higher the manifold pressure, the more the valve restricted the return flow and the richer the mixture became.

BAE took a slightly different approach with the fuel-injected Datsun 280Z. They added a device that opens the cold-start circuit when manifold pressure goes above ambient, causing the engine to run richer. In this instance, the additional fuel serves as an anti-detonant when the engine is being operated above boost pressures of 2 psi.

Janspeed used still another approach on its 280Z system by adding two Bosch cold-start valves from a BMW. One valve starts operating at about 2-psi boost. The second opens at about 5 psi, giving a progressive enrichment with increasing manifold pressure.

Engines equipped with Bosch K- and L-Jetronic fuel injection use an air flow sensor to judge engine fuel requirements. One way to enrich these engines is to use a blowoff valve on the intake manifold—rather than an exhaust wastegate—to control boost pressure.

The blowoff valve will control boost pressure and at the same time cause the engine to run richer. The air flow sensor—which is ahead of the blowoff valve—signals for a certain fuel/air ratio based on *incoming* air flow. When some of the air is bled off through the blowoff valve, the fuel/air mixture entering the cylinders becomes richer.

When using a blowoff valve to enrich the mixture, mount the valve *upstream* of the cold-start valve. Otherwise, a malfunctioning cold-start valve could cause the blowoff valve to release a combustible fuel/air mixture into the engine compartment.

EVAPORATIVE-CANISTER PROBLEMS

One of the modifications made to carburetors to reduce evaporative emissions is a vent from the fuel bowl to a charcoal canister. The canister is usually connected to the fuel tank to capture gasoline fumes from there.

Once again, the inventors of this system did not have turbocharged cars in mind—or at least those that blow through the carburetor. Pressure from the compressor will pressurize the float bowl, canister and fuel tank.

A check valve placed in the vent line to the charcoal canister will keep pressure out of the canister and fuel tank. It should open at 1/4-psi pressure. A line from the check valve into the air cleaner bleeds off pressure.

CARBURETOR CAPACITY

In general, a high-output turbocharged engine requires the same or even less carburetor capacity than a high-output naturally aspirated engine. For example, a normally aspirated 1971 Oldsmobile 350-CID engine used a 725-cfm Quadrajet. When Rajay turbocharged the car, they got best results with a single Holley 600-cfm four-barrel.

Crown Manufacturing's Datsun 240Z installation used only one of the two SU/Hitachi carburetors provided on the engine, yet achieved near 100-mph performance in the 1/4 mile.

AIR CLEANER

Some people think an air cleaner is a waste of time and power on a high-performance engine. Ted Trevor says, "There may be instances in which an air cleaner is not desired, like a boat that never comes to port or an airplane that never lands. All other applications of internal combustion engines need air cleaners."

The compressor impeller in Figure 6-20 is a good example of what can happen when an air cleaner is not used. A small rock or a piece of metal, such as a screw or nut, will wipe out a compressor impeller. Parts of the impeller will then pass through the engine, perhaps causing extreme damage.

With this in mind, carburetor-butterfly screws should be staked in place to prevent them from backing out. This is normally done by the carburetor manufacturer, but check to make sure. If it has not been done, do

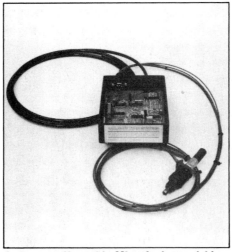

Callaway offers this Microfueler enriching device. Microprocessor controls solenoid-type fuel injector using outputs that sense engine fuel requirements. Location of injector in manifold is critical to even fuel distribution. Photo by C. E. Green.

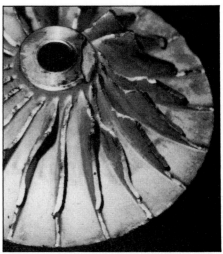

Figure 6-20—Compressor impeller with chopped-up blades. This was the result of foreign matter being drawn into the compressor.

Allen Osborne made use of Hobbs switches and cold-start fuel-enrichment valves to supply additional fuel required by his turbo-Datsun L28 engine. As manifold pressure increases, each switch opens a valve in staged increments of 5 psi. Although fuel-distribution was a problem, car set a record two-way average of 188.394 mph at Bonneville. Direct-port fuel injectors have since solved the distribution problem. Photo by Tom Monroe.

it! Use a backup anvil on the opposite side of the shaft to prevent bending the shaft.

7 Ignition

McLaren Engines developed this turbocharged 209-CID stock-block Buick V6 for Indy Cars. Warner-Ishi RHD9 turbocharger pumps 57-in.Hg pressure for an engine output of over 750 HP. Computer-controlled electronic ignition system is used. Note crank-position pickup at crankshaft nose and electronic "distributor" between injectors.

The ignition requirements for a turbocharged engine are not much different than for a high-performance naturally aspirated engine. The ignition system has to make a spark at the right time to fire the charge. What goes on outside the combustion chamber is similar in both cases. Higher combustion-chamber pressure creates a few extra problems for the turbocharged engine.

IGNITION-SYSTEM BASICS

The difference between a conventional breaker-points ignition and the transistor-switched type is how the coil is fired.

Conventional—With the conventional type of ignition, whenever the points are closed, the current flows through the primary windings in the spark coil. This current builds a magnetic field in the coil.

When the points open, the magnetic field collapses. The primary winding of the coil loses its ground and the current must travel through the secondary winding to ground—the spark plug. The low-voltage current in the primary winding is changed to high voltage in the secondary winding to fire the spark plug.

The efficiency of the system depends a lot on how much time the coil has to store current. The faster the engine runs, the less time the coil has to build up a charge. If the engine speed gets too high, the secondary voltage will be insufficient to fire the spark plug.

One solution to this problem is to increase the voltage/amperage to the coil. Unfortunately, mechanical contact points can't handle it. Increasing voltage or amperage to the coil to give a hotter spark will overload and burn the contact points.

Electronic—In an electronic-ignition system, the triggering device is magnetic, optical or mechanical. Starting and stopping the current flow is accomplished by a transistor.

With no points to burn, current flow is not limited to the low voltage required with the point-type ignition. Consequently, a higher-amperage current can be fed to the coil. This results in a higher-voltage, or "hotter," spark.

Some of these systems step up the battery voltage before it enters the coil. The higher voltage means less time is needed to build up a charge in the coil.

Some turbocharged engines running at high speed and high boost will suddenly backfire for no apparent reason. This can be a symptom caused by a too-lean mixture. It can also result from not enough secondary voltage. A good electronic ignition system is sometimes the answer when this problem occurs.

Spark Plugs—Most spark-plug cleaners include a small chamber in the machine in which the spark plug can be tested. The plug is inserted and air pressure applied to see if the plug will fire under pressure. This chamber has a window so the actual spark may be observed.

A little experimentation will show that as the pressure in the chamber increases, the gap on the spark plug must be decreased to ensure it will fire.

This same problem exists in a turbocharged engine. The combustion-chamber pressure in a turbocharged engine is considerably higher than in a naturally aspirated engine with the same compression ratio. Therefore, the plugs will require a narrower gap or higher voltage to ensure they'll fire.

If you're using a conventional point-type ignition, the spark gap should be reduced to aid ignition under maximum boost. The recommended gap on most passenger cars with conventional ignition is 0.035 inch. It is a good idea to reduce this to 0.025 inch when adding a turbocharger.

Reducing plug gap is not necessary on the original-equipment high-energy ignitions, like the General Motors HEI or Ford DuraSpark. These use a wider plug gap—up to 0.080 inch—and can handle the in-

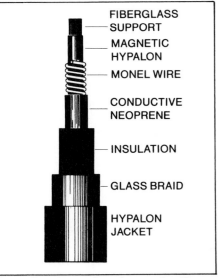

Magnetic-suppression wire construction details

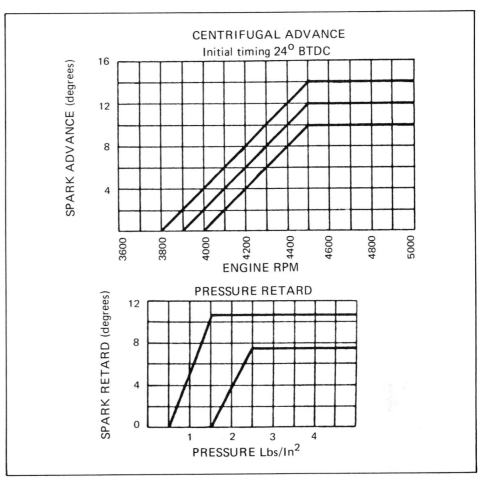

Corvair Spyder distributors used a pressure-retard unit instead of vacuum retard. Spark retards when manifold sees boost, partially opposing the affect of centrifugal advance. Initial distributor setting is 24° BTDC. Curves show advance/retard production tolerances of centrifugal-advance and pressure-retard mechanisms.

creased combustion-chamber pressure of turbochargers.

Plug heat range is determined by the use of the engine—just as it is in a naturally aspirated engine. However, even in ordinary street use you should use one heat range colder than originally recommended after the engine is turbocharged. Don't use too cold a plug or the turbocharged engine will develop the same plug-fouling problems the naturally aspirated engine would.

Spark-Plug Wires—Most ignition wires consist of cotton or similar material impregnated with carbon. Used to reduce radio and TV interference, these wires are called *TVR*.

On a typical passenger car, in which the plugs are changed about once a year, TVR cables might last the life of the car. In the hands of an enthusiast it's a different story. The whole harness might be removed and replaced dozens of times. Flexing the cables can break the continuity and increase the resistance.

In one instance, an engine was missing badly for no apparent reason. Checking the resistance of the spark-plug leads, we found one lead measured 200,000 ohms; it should have been about 20,000 ohms. The wiring harness was replaced and the engine ran without missing.

I should also note that TVR wire cannot be used with capacitive-discharge ignition systems. These systems will quickly destroy TVR wire.

TVR-type wires should be replaced with steel-core or Magnetic-Suppression Wire (MSW). MSW spark-plug leads should be installed if the vehicle is radio-equipped or if radio/TV interference is a concern. MSW has all of the advantages of the TVR wire with none of the disadvantages.

SPARK CONTROL

Ignition systems on present-day passenger cars are equipped with several mechanisms to control spark advance. Advance is restricted any time it might increase NO_x emissions.

The mechanisms vary from car to car. In general, most engines have no vacuum advance in the lower gears, no vacuum advance until the engine has come up to temperature and no vacuum advance before a preset engine speed. Some also have spark retard during deceleration and at engine idle.

As bad as it all sounds, a turbocharger may be installed in an engine with these ignition devices and still get very good performance. *A turbocharger is one of the few ways of getting a great increase in performance without disturbing the emission-control devices.*

An engine not equipped with emission devices should have its spark curve adjusted to give maximum advance without detonation. This should be achieved for both the naturally aspirated and turbocharged modes.

Pressure Retard—When the Corvair was turbocharged, high-octane gasoline was available. A turbocharged engine is normally a low-compression engine. If high-octane gasoline is used, the engine can tolerate extremely advanced timing at the lower speeds when no turbocharging takes place.

Unfortunately, the days when you could buy high-octane gasoline at the local gas station are gone. Using lots of spark advance prior to boost may only cause detonation.

The Corvair Spyder engine was set up with a lot of advance at lower speeds. Initial timing was set at 24° BTDC, which gave pretty good performance until about 2000 rpm. At this speed the intake-manifold pressure became positive.

Pressure retard was used on the distributor instead of vacuum advance. When boost reached 2 psig a pressure-retard diaphragm on the distributor retarded the spark approximately 10 crankshaft degrees.

At 3800 rpm the engine could stand more spark advance despite the high intake-manifold pressure. Centrifugal advance started at this speed and advanced the spark another 12° by 4500 rpm. Liquid-cooled engines do not seem to be as sensitive to spark timing and can usually stand more advance at less than 3800 rpm.

The combination of centrifugal advance and pressure retard was developed on the dynamometer. At the time it seemed the best compromise for maximum power without detonation.

As mentioned in the carburetion chapter, the Corvair carburetor had no provision to add extra fuel under boost. A fuel-enriching device or water or water/alcohol injection would have eliminated the need for the pressure-retard device.

If you choose to use a pressure retard, you may be able to find the Corvair diaphragm. The pressure-retard diaphragm from the Corvair Spyder fits other four- and six-cylinder GM distributors from 1962—74. It can be adapted to other distributors.

The centrifugal-advance/pressure-retard curve for the Corvair is not the best for another engine. The best spark curve should be worked out for each individual engine. This rule also applies to switching components in the same engine, such as the camshaft. The distributor must be re-curved if the optimum spark-advance curve is to be found. This can be done either on a dynamometer or on the road.

For example, a 1600cc Volkswagen engine seems to operate best with 26° total advance at 4000 rpm. Advancing the spark beyond this cools the exhaust temperature. The engine then

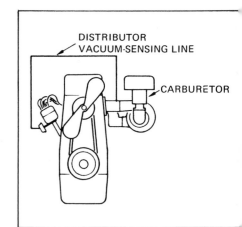

Figure 7-1—Vacuum-advance sensing-line location

produces lower boost pressure and less horsepower.

With less than 26° spark advance, the Volkswagen gained boost pressure but lost power due to the retarded timing. Retarded timing also increased head temperature.

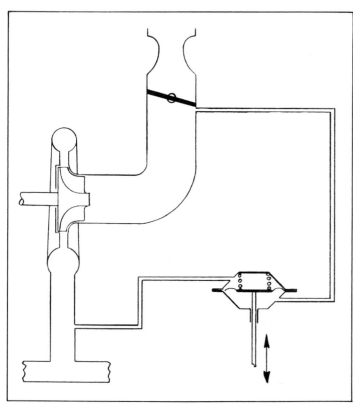

Figure 7-2—Bill Reiste's double-acting diaphragm

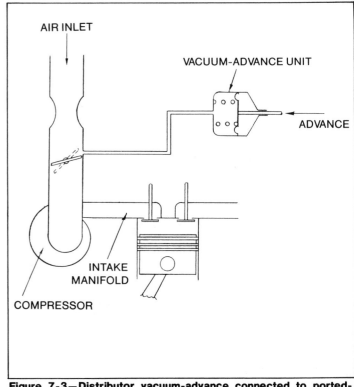

Figure 7-3—Distributor vacuum-advance connected to ported-vacuum source at carburetor

Vacuum Advance—Good results have been obtained using fuel enrichment or water or water/alcohol injection with a vacuum-advance device. The vacuum line should be attached to the intake manifold. Here it will sense pressure in the manifold rather than at the spark-advance port, Figure 7-1. Connecting the hose to the spark-advance port could give too much advance under boost and cause detonation.

Bill Reiste has had good success with some of the dual-diaphragm vacuum-advance/-retard mechanisms used on emission-controlled engines. These are used primarily on '72- and-later Ford products and certain Bosch distributors.

He connects one port to the intake manifold and the other to the carburetor, Figure 7-2. The different combinations of boost and vacuum give retard at idle, advance at part load and retard when supercharged.

One of the problems with connecting a standard vacuum advance to the intake manifold is that it gives full advance at idle. This creates a bog just off idle.

Leaving the advance connected to the carburetor port just above the butterfly is fine at wide-open throttle. But a problem occurs under boost at part throttle; timing remains advanced when it should retard, Figure 7-3.

Andronico Gonzales solved this problem by adding a check valve and orifice, Figure 7-4. With this system, the distributor retards at idle, advances at part load and retards again under heavy load or boost. The engine will have the proper advance, regardless of the butterfly position.

As in the case of naturally aspirated engines, there is no "best" ignition timing for all turbocharged engines. To repeat: The ideal timing for any engine will be the result of trial-and-error testing on the dyno or road.

Electronic Timing Controls—Introduced in 1978, Figure 7-5, Buick's approach to ignition-timing control on their turbocharged V6 is the forerunner of more sophisticated systems that are in use today and those to come.

The detonation sensor "hears" detonation and retards the timing. It reacts so quickly that you can usually hear only one or two detonation "rattles" before the Electronic Spark Control (ESC) module has retarded the ignition and completely eliminated detonation.

The ESC approach allows using near-optimum spark advance at all times. Detonation is effectively controlled, even though gasoline octane may vary considerably.

This is a good step toward obtaining best economy. It also eliminates the possibility of an unaware owner burying his foot in the throttle and leaving it there while the engine is destroyed by detonation.

This system has been made possible by the rapid advances in electronics technology. Modern ignition and carburetion or fuel-injection systems are controlled by devices sensing throttle position, load, speed, detonation, oxygen, air/fuel ratio and anything else pertinent to combustion.

On an experimental basis, electronic engine controls on the '83 Buick V6 Indy pace car controlled the turbocharger wastegate to drop boost pressure and retard the spark for detonation control. Saab has also used waste-gate control successfully. Many similar systems are in the works.

SUMMARY

Ignition does not pose any more difficult problems on a turbocharged engine than on a naturally aspirated engine. Timing, however, is quite different. Enough effort should be spent on the advance curve or the results of turbocharging your engine might be disappointing.

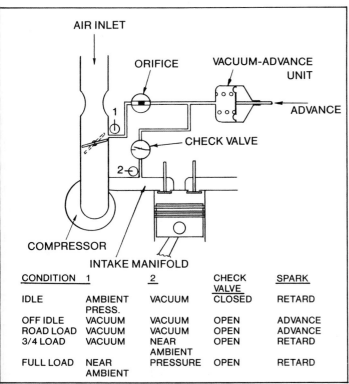

CONDITION	1	2	CHECK VALVE	SPARK
IDLE	AMBIENT PRESS.	VACUUM	CLOSED	RETARD
OFF IDLE	VACUUM	VACUUM	OPEN	ADVANCE
ROAD LOAD	VACUUM	VACUUM	OPEN	ADVANCE
3/4 LOAD	VACUUM	NEAR AMBIENT	OPEN	RETARD
FULL LOAD	NEAR AMBIENT	NEAR AMBIENT PRESSURE	OPEN	RETARD

Figure 7-4—Modification to system prevents advance at part throttle when intake manifold has little or no vacuum.

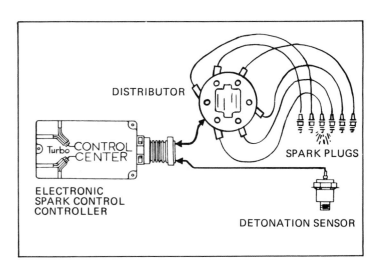

Figure 7-5—Introduced on the Buick V6, detonation control is accomplished by Electronic Spark Control (ESC). Signal is received from detonation sensor on the intake manifold. When detonation occurs, controller retards spark electronically up to 22° at the crankshaft. Unit readvances spark when detonation ceases. This system is used on virtually all new GM cars using Computer Command Control.

8 Exhaust

Stripped Chaparral Indy Car shows tightly packaged exhaust system that allows for wider ground-effects tunnels on underside of car. Exhaust system must be durable and enhance engine power, just as if engine were normally aspirated. Photo courtesy of Pennzoil.

On a naturally aspirated passenger-car engine, the exhaust system performs two functions: It carries combustion products from the engine to the rear of the vehicle and reduces the noise level. At the same time it is desirable to have as little exhaust-system back pressure as possible.

On a racing car, exhaust pipes from the individual cylinders are sometimes cut to a certain length. These *primary* pipes are usually brought together in a *collector.* The scheme is to use normal exhaust pulses to assist the flow of exhaust gases from the engine.

In either of the above cases, the exhaust system should not only rob the engine of as little power as possible, it should enhance power.

Turbocharged-engine exhaust systems are different. They duct the hot, high-pressure, high-velocity gases

from the engine to the turbocharger. This temperature, pressure and velocity should be maintained as much as possible.

As mentioned in the turbocharger-design chapter, a turbine housing increases exhaust-gas velocity to about 2000 feet per second. This increase in velocity is necessary to maintain turbine speed. For example, the tip speed of a 3-inch-diameter turbine wheel is about 1600 feet per second at 120,000 rpm.

Any velocity or pressure lost in the exhaust system must be regained in the turbine housing. Let's say the exhaust gas comes out of the exhaust port of the engine at about 300 feet per second. Not much is gained if it is slowed to 100 feet per second, then is speeded up to 2000 feet per second in the turbine housing.

For this reason, *an extremely large-diameter exhaust manifold is not recommended.* The exhaust-port area is about right for the cross-sectional area of the manifold runners.

Smooth-flowing exhaust headers are desirable, but are difficult to fit into the tight engine compartment of a road car. Consequently, turbocharged engines in road cars have relatively simple, somewhat restrictive exhaust systems.

One of the drawbacks of most V8 engines is they have 90° crankshafts. This causes two adjacent cylinders on each bank to fire 90° apart, causing instantaneous back pressure on the cylinder that fires next. This overlap of exhaust pulses is not critical on a naturally aspirated engine unless it is used for all-out racing. Maximum performance requires ducting the gases

Katech-prepared turbocharged R5 Renault produces 263 HP at at 7200 rpm. This 1100cc pushrod engine is equipped with Bosch fuel injection, intercooler and dyno-developed equal-length primary exhaust pipes. Exhaust headers improved power and throttle response.

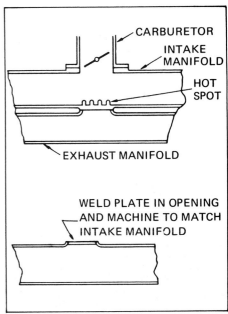

Figure 8-2—One way to close off in-line engine heat riser

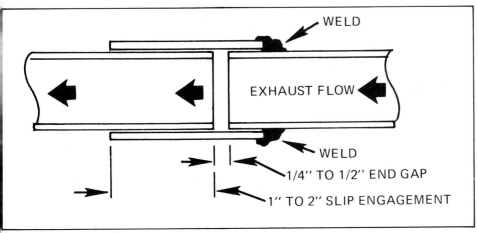

Figure 8-1—Typical slip joint for exhaust

separately from the two ports rather than through a *log*-type manifold that fits tightly against the cylinder head.

The problem is easily solved by using a 180° crankshaft if the vibration can be tolerated—and the cost of having such a crankshaft custom fabricated.

Slip Joints—Some engines, particularly V6s and V8s with one turbocharger, require long exhaust pipes upstream of the turbocharger. This creates two problems: Long exhaust pipes mean additional heat expansion and higher-than-normal stress applied to the pipes. And, the pipes from each head must be joined before the turbocharger. On these applications, Don Hubbard suggests

using a slip joint in the exhaust duct between the two heads, Figure 8-1.

Slip joints have been used on aircraft engines for many years. There's surprisingly little leakage. The inside pipe runs hotter than the outside, causing the joint to tighten after heat expansion takes place. This relieves much of the stress that would otherwise occur with a rigid joint.

Don suggests 0.004 to 0.005-inch diametral clearance for mild steel and 0.005 to 0.008-inch for stainless steel.

Insulating the Exhaust System—It is not uncommon to insulate an exhaust pipe on a turbocharged engine to keep the exhaust gases as hot as possible. However, insulation can cause problems.

When a pipe is insulated, it becomes considerably hotter. It may even get hot enough to burn out. Even if the pipe does not burn out, it tends to expand more because of the higher temperature.

If the ends of the pipe are restrained, this expansion puts additional stress on any bends and joints and may cause the pipe to crack. This problem is particularly evident with an air-cooled-Volkswagen exhaust. Again, a slip joint will help solve this problem.

Exhaust-Manifold Gasket—Many passenger-car engines do not have gaskets between the exhaust manifold and the head. Normally, this is not important as far as engine power is concerned, although a leak could be dangerous to the passengers.

A small leak at this joint on a turbocharged engine will cause a considerable boost loss. The effect is compounded because the turbine will slow, which will slow the compressor. The compressor will then develop less pressure, less pressure will slow the turbine, which will . . . . More is said about exhaust leaks at the end of this chapter.

If available, a good sandwich-type metal/asbestos gasket should be used. Next best is a gasket cut from asbestos material backed with perforated metal. Unsupported asbestos will blow out in a short time.

Beaded stainless-steel gaskets work

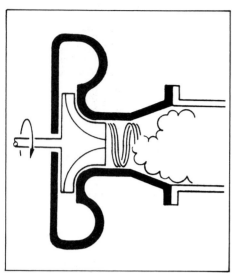

Figure 8-3—Turbine discharge with rapid expansion converts swirl to turbulent flow

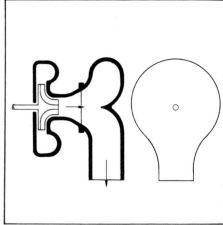

Figure 8-4—Banjo-type turbine discharge

well if both surfaces are very flat. But they should not be used with plate-steel exhaust-header flanges regardless of surface flatness. These type flanges warp.

Even the best gaskets will not work if the exhaust manifold is warped. Check the mating surfaces and if they are warped have them machined flat.

Heat Risers & Carburetor Heaters—When two or more turbochargers are used on an engine, best results are obtained when all the turbochargers put out the same boost. For this reason, some type of balance tube should be used between the two manifolds to equalize boost pressure from side to side.

A V-type engine ordinarily has a heat-riser passage through the intake manifold to prevent fuel condensation and carburetor icing. Leaving the heat riser open will ensure equal pressures in both exhaust manifolds. If there is no heat riser, join the exhaust manifolds with a balance tube at least 3/4 inch in diameter.

In-line engines with intake and exhaust manifolds on the same side ordinarily have an opening in the exhaust manifold under the intake manifold. The opening creates a "hot spot" in the intake manifold. This performs the same function as a V-type heat riser.

The joint between the exhaust and intake manifold may not be perfect and may leak exhaust gas if used on a turbocharged engine. The opening can be blocked by welding a plate in place, Figure 8-2. The plate should be

machined flat so it will not interfere with the intake manifold.

TURBULENT FLOW IS HELPFUL

Exhaust gases exiting the turbocharger will be spiraling like a helix, Figure 8-3. The direction of this spiraling can be either the same or opposite turbine-wheel rotation. Which direction depends on exhaust-gas speed.

The turbocharger is used over a broad range of engine speed and power. Consequently, the exhaust gases will sometimes be going faster than the turbine-wheel exducer—and sometimes slower.

When the gases are going faster than the turbine wheel, they will rotate in the opposite direction from the turbine rotation. When they are going slower than the turbine wheel they rotate in the same direction as the turbine.

In either case, the path for the exhaust gases is considerably longer than if they were coming out axially. For this reason, it is desirable to break up the swirl and change the gases to turbulent flow. This should be done as soon as possible after the exhaust leaves the turbine housing.

One way to do this is to have a sharp diffuser angle on the turbine housing, Figure 8-3. This requires a large-diameter exhaust pipe from the turbine housing. Once turbulent flow is established, usually after 18 to 24 inches, exhaust-pipe diameter can be reduced.

A smooth reduction in the exhaust-pipe cross section will not create much additional back pressure and will help quiet exhaust noise. In some cases it will do such a good job that a muffler will not be required.

When a muffler is used, don't be misled by the amount of noise you hear at the back of the vehicle. The original Corvair Spyder muffler was extremely quiet but added very little back pressure. When comparing mufflers it is best to monitor pressure using a gage between the turbine exhaust and the muffler.

Banjo Turbine Discharges—Sometimes turbocharger installations end up with the exhaust very close to the fire wall or some other obstacle. When this happens, there may not be enough room to get an elbow between the turbine outlet and the exhaust

pipe. A banjo-type turbine discharge, Figure 8-4, can be used for routing the exhaust in close quarters.

Bristol cars ran into this problem with their turbocharged Beaufighter which uses a Chrysler 360-CID engine. This car has close clearance between the turbine discharge and the engine-compartment fire wall.

Dennis Sevier, Chief Engineer of Bristol, said a short-radius elbow was tried but the car lost power on acceleration and top speed. He replaced the elbow with a banjo-type fitting and found it worked at least as well as a straight exhaust.

This car accelerates from 0 to 100 mph in 13.8 seconds—not bad for a four-passenger car with a family-car-size trunk.

Actual exhaust-pipe diameter on a turbocharger system often depends on the application. Even on an all-out racing engine it may not be desirable to use as large an exhaust pipe as possible.

A slight restriction in the exhaust pipe can be an effective high-speed boost control. An exhaust restriction can be used as a fail-safe method to limit manifold pressure without low- or medium-rpm power penalty.

EXHAUST LEAKS COST POWER

An exhaust leak between the engine and turbine can be aggravating because the turbocharger will not produce full boost and maximum power.

A leak large enough to cause considerable loss of boost may not be detectable to the ear. It will no

Banjo-type turbine discharge for use in close quarters

Bristol Beaufighter engine compartment. Car has Rolls luxury and Porsche performance, with ample room for four people. The turbocharged 360-CID Chrysler pushes car to 0-30 mph in 2.4 seconds, 0-60 in 5.8 seconds and 0-100 in 13.8 seconds.

necessarily leak at idle conditions. At high rpm everything makes so much noise that even a large leak between the engine and turbine is hard to find.

The easiest way to check for a leak is to block off the turbocharger exhaust completely while the engine is idling. If the engine continues to run, the leak is too big. When a turbocharged engine does not deliver full power, the turbocharger is always blamed. If the turbine and compressor wheels are intact and the shaft can be rotated, chances are 100 to 1 there's a leak in the exhaust system leading to the turbo.

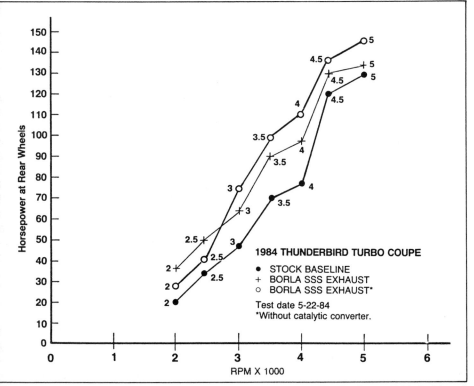

1984 THUNDERBIRD TURBO COUPE

- ● STOCK BASELINE
- + BORLA SSS EXHAUST
- ○ BORLA SSS EXHAUST*

Test date 5-22-84
*Without catalytic converter.

(chart axes: Horsepower at Rear Wheels vs. RPM X 1000)

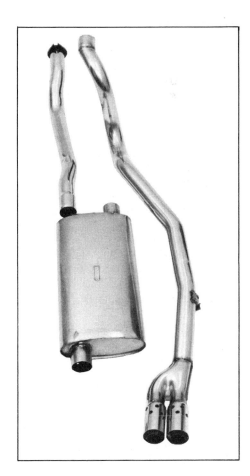

Chassis dynamometer curves of Borla exhaust-system-equipped Turbo Coupe T-Bird with and without catalytic converter versus stock system: Significant powers gains are realized over full rpm range. Exhaust system, shown at right, bolts in place of production unit.

9 Lubrication

Janspeed-turbocharged 308GT Ferrari. Design of engine dry-sump oil-system required separate turbo oil system. Turbocharger oil pump—under turbocharger oil filter (1)—is driven off end of camshaft. Turbocharger oil is passed through its own cooler (2). Five-liter oil reservoir is filled on far side of engine (3).

Turbocharger bearing and seal design was covered in Chapter 2, Turbocharger Design, but that is only half of the story. It is essential to provide clean oil to the bearings, and just as important to get the oil back into the crankcase.

Oil Type—All turbochargers described in this book are designed to use engine lubricating oil. The actual type and viscosity is dictated by what is required for the engine. In general, a turbocharger will operate on any oil that works in an engine as long as it is clean.

Oil Filters—It is not necessary to install a special oil filter for the turbocharger if the engine is equipped with an oil filter that stops particles larger than 30 microns. If the engine is not equipped with a full-flow oil filter, then a separate filter for the turbo is definitely recommended.

The filter should be placed in the oil-inlet line. Use the type with a built-in bypass, so oil will still get to the turbocharger even if the filter is clogged. It is not uncommon for a turbocharger to fail from lack of oil because the user didn't service the filter. A turbocharger running on dirty oil will last only a few hours.

WHERE TO GET OIL FOR TURBO—HOW MUCH IS NEEDED?

It is usually safe to pick up oil for the turbocharger where the oil-pressure-light or -gage sender installs. Put a tee in the line at the turbocharger end and *reinstall the sender.* Any restriction at the point where the oil is picked up will show up as low oil pressure as soon as the engine is started.

Another test should be made to ensure adequate oil flow to the tur-

bocharger by disconnecting the oil drain *from* the turbocharger while the engine is idling. Measure oil flow by running it into a bucket for one minute. Don't let the engine run any longer. Otherwise your engine's lubrication system will pump itself dry. *Be sure to add oil to the engine after doing this test.*

A 3-inch turbocharger requires at least one-half gallon of oil per minute. About one gallon per minute is required by a 3.5- or 4-inch turbocharger. On multiple-turbo installations, be sure to check *each* turbocharger.

For a 3-inch turbocharger, 1/4-inch-OD tubing is sufficient. Use at least 5/16-inch-OD tubing for a 3.5- or 4-inch turbocharger. These tubing sizes are adequate if the engine is running on hot oil with at least 30-psi oil pressure.

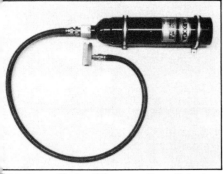

Accusump Turbo Oiler supplies oil to turbocharger immediately after engine is shut off. This helps prevent shaft and bearing damage and oil coking during hot, high-rpm shutdown.

If oil viscosity is too high for ambient temperature conditions, no oil will flow to the turbocharger when the engine is first started. Here again, the oil-pressure light or pressure-gage connection at the turbocharger will show this. Starting the engine and driving off before oil gets to the turbocharger can wipe out the bearings in minutes.

Never use a screen or restrictor orifice in the turbocharger oil-supply line. Either will clog quickly and cause early turbocharger failure. If the oil pressure is so high that the added flow causes drainage problems, place a pressure-reducing valve in the line, not an orifice.

Oil Pressure—When an engine is shut off, oil pressure drops to zero. This is no problem for the engine because it and the oil pump stop simultaneously. The story is different for a turbocharger on the engine, particularly if it's a race engine.

Imagine this extreme condition: A race car with a turbocharged engine charges into the pits after several 150-mph-average laps, brakes hard to a stop and the engine is shut off. The engine and turbo are very hot and the turbocharger is spinning at 100,000 rpm without oil pressure to the bearings. The result is high bearing wear or, worse yet, bearing seizure.

This can be solved with an *oil accumulator* teed into the the turbocharger oil-supply line. Charged with oil under pressure from the engine-lubrication system, the accumulator will maintain oil pressure at the turbocharger bearing for a short time after the engine is shut off; long enough to prevent bearing damage. Such an oil accumulator—the Accusump III Turbo Oiler—is available from Mecca Development, Inc.,

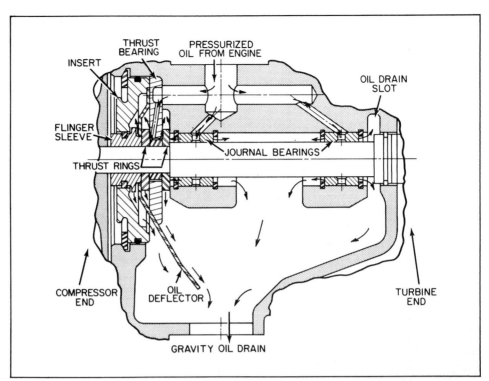

Oil-flow path through turbocharger: Setup uses piston rings to seal oil at compressor- and turbine-ends of shaft. Note oil deflector and large area for oil drain. Drawing courtesy of Schwitzer.

Oil-pressure sender in turbocharger oil line monitors engine- and turbocharger-oil pressure.

Route 41, Sharon, CT 06069 or Auto World, 701 N. Keyser Ave., Scranton, PA 18508.

Oil Return—Oil entering the turbocharger is relatively air free. As oil passes through the bearings with the turbo running as high as 130,000 rpm, air is whipped in and the oil looks like dirty whipped cream. For this reason, use a drain line from the turbocharger that is much larger than the oil-supply line.

The line must angle downward at all points and have no kinks or "sink traps." The drain line must dump oil to the crankcase *above the oil level.*

The sketch at the top of page 76 shows the oil line entering the crank-

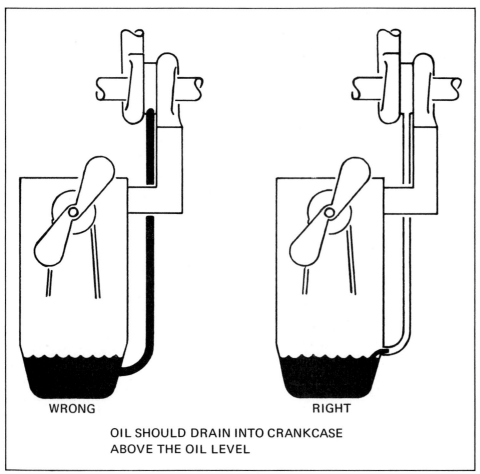

WRONG

RIGHT

OIL SHOULD DRAIN INTO CRANKCASE
ABOVE THE OIL LEVEL

Figure 9-1—Right and wrong way of draining oil from turbocharger

Large-diameter drain hose routes turbocharger drain oil straight down to top of engine sump on this GIK Mercedes 220/240D installation.

case below the oil level. This causes foamy oil to build up in the line and back up into the turbocharger bearing housing. The only place it can go from there is out through the seals.

Many people have torn down turbochargers to replace leaking seals only to find that they appeared as good as new. Chances are they *were* as good as new.

Before the days of emission controls, the crankcase was equipped with a breather to prevent pressure buildup in the crankcase. The pressure is caused by blowby, which occurs in all engines. Blowby is the result of high-pressure gases leaking past the piston rings of the engine. The crankcase breather allowed the blowby gases to escape into the atmosphere. A filter element in the breather prevented dirt from entering the engine.

On a turbocharged engine, it is especially important to service these breathers regularly. If they become partially clogged, pressure will build in the crankcase, preventing a free flow of oil draining from the turbocharger.

All passenger-car engines built today are equipped with a positive crankcase ventilation (PCV) system. The design varies with the engine, but usually a hose connects the valve covers to the air cleaner. Another hose runs from the valve covers to the intake manifold.

The line to the intake manifold is normally equipped with a PCV valve. This valve restricts flow from the crankcase when intake-manifold vacuum is high and blowby is low. It has low restriction at full power when intake-manifold vacuum is low and blowby is high.

If the PCV valve is not serviced regularly, it will stick and cause pressure to build up in the crankcase. This will prevent the oil from draining properly from the turbocharger, and will probably cause leakage through the seals.

If your turbocharger shows signs of oil leaking into either the turbine or compressor housing, be sure to check all these points. It's frustrating to tear down the turbocharger only to find that the problem is elsewhere.

Drain Location—When installing a turbocharger—particularly if it is part of a bolt-on kit—most people do not want to remove the engine. Unfortunately, attaching the oil drain o the pan is difficult unless the pan is removed. In many cases, the pan can be removed only by removing the engine. Because of the problems with mounting an oil-pan drain, some other place to drain the oil is desirable.

If the turbocharger or turbochargers are mounted high enough, the oil may be drained into one or both of the valve covers.

Valve-cover drains don't work on all engines. Some engines have a hard enough time draining the small amount of oil used to lubricate the rocker arms and valves. Additional oil from the turbochargers can flood the rocker area and back up into the turbocharger. This is true of the small-block Chevy V8.

On overhead-valve in-line engines, the side plate covering the pushrods is often a good place to drain the turbocharger oil. Some V-type engines have a hole through the intake manifold into the valley between the heads.

Other V-type engines may have a place in the intake manifold you can drill to drain oil into the valley. Before you drill holes in the manifold, be sure you won't hit an intake runner, heat-riser passage or water passage.

If the turbocharger is mounted so low it is not practical to drain the oil back to the engine, a scavenge pump must be used. This is often the case on an airplane engine, where the only practical place to mount the turbocharger is beneath the engine.

The scavenge pump must have a much greater capacity than the amount of oil used by the turbocharger. Again, this is due to the air that gets mixed with the oil. A scavenge pump for a 3-inch turbocharger should have a capacity of 1.5 gallons per minute.

Aviad Metal Products Company and Weaver Brothers both manufacture belt-driven scavenge pumps specifically designed for turbocharged engines. Martin Schneider also markets an electric pump for the same application. These units suck the oil out of the turbocharger and pump it into the engine sump.

High-Capacity Oil Pumps—Most engines have an oil pump with enough capacity to handle the addition of a turbocharger. As I mentioned, low oil pressure at the turbocharger is probably due to a restriction at the oil source rather than insufficient oil-pump capacity.

Another GIK installation routes drain oil to banjo fitting at front cover on this Volvo.

If there is no restriction on the oil supply and the oil pressure is still low, then you must use a larger-capacity oil pump. High-capacity pumps are available for many engines, but chances are one won't be necessary.

Innumerable high-capacity oil pumps have been advertised as a way to extend crankshaft life on VW engines. Yet turbocharged Volkswagens have gone long periods without excessive crankshaft wear. Many have gone what would be considered a high number of miles on a naturally aspirated engine. These engines were equipped with an oil filter, which is more important than a high-capacity oil pump. Therefore, consider a filter a must on a turbocharged VW.

Oil Coolers—A turbocharger running at design speed will add about 80F to the oil as it passes through the bearings. On a passenger car, the turbocharger is idling most of the time. The additional heat occurs only occasionally. Typically the oil does not run much hotter than on a naturally aspirated engine.

This is not necessarily true on a truck, bus or a motor home. Here the turbocharger may be used for extended periods. On any of these applications I recommend installing an oil-temperature gage.

For installations requiring turbocharger to be mounted lower than engine sump, draining oil is a problem. Martin Schneider uses this electric pump to return turbocharger oil to sump.

If the oil temperature consistently goes above 250F, install an oil-cooler. These come in many shapes and sizes, depending on the engine and the application. The important thing is to keep the oil cool to prevent oil breakdown and oxidation.

SUMMARY

A turbocharger lubricated with clean oil at engine pressure can last many years without visible signs of wear on the bearing journals. But if dirty oil, or no oil at all, is supplied to the turbocharger—even for a short period of time—the chances are the unit will be short-lived.

10 Controls

Turbocharger boost on Indy Car can be controlled from cockpit. Importance of turbocharger boost is illustrated by position of boost gage and control knob; gage is mounted above steering wheel and control knob is just to left of steering wheel. Quick-release steering wheel has been removed. Photo by Tom Monroe.

Turbo Performance Products' neat installation in Porsche 912. Boost is controlled by an IMPCO valve. Oil cooler is attached to engine cover. Waste gates are preferable to IMPCO spoiler-type control because they do not restrict the intake, so there is no heating of the fuel/air mixture.

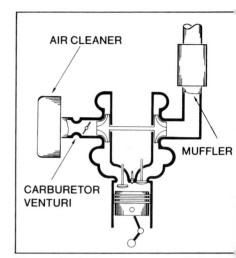

Figure 10-1—Boost pressure limited b fixed restrictions

Given enough time and money, is possible to achieve optimum perfor mance for any turbocharger/engin combination without any movabl controls, Figure 10-1. It's also possibl to match this *free-floating* combinatio to all operating conditions—assumin they are known. The problem is mos of us don't have the time or th money to do it.

WHY A CONTROL?

Unless some type of control i used, the typical turbocharged engin is a compromise. Either it will no achieve maximum horsepower or i won't have peak response. For stree use, a turbocharger can be matched t an engine without controls and sti give excellent performance. This i perhaps true even for a Bonneville type racer, where top speed is mos important.

This is not necessarily true of a engine for track, drag or boat racing For example, when a driver is qualify ing his car, he must get maximun horsepower and may not be particular ly concerned with durability. Durin the race, horsepower and durabilit are equally important.

If an adjustable control were no used here, it would probably be neces sary to use a different turbocharge for qualifying and racing. Boost migh be controlled by changing just the tur bine housing. But even here, it's no

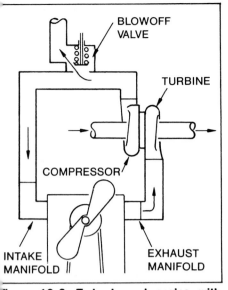

Figure 10-3—Turbocharged engine with compressor-blowoff valve

practical to change turbine-housing sizes after each run.

Another advantage of a turbocharger control is that it allows the turbocharger to be run at or near its maximum speed.

TYPES OF CONTROLS

Turbocharger controls are generally divided into two categories: those that limit turbocharger speed and those that limit compressor-outlet pressure, or boost. Controls that limit turbocharger speed prevent the turbocharger from destroying itself. Those that limit boost keep the turbo from destroying the engine.

Modern turbochargers can produce more pressure than any engine can stand. Consequently, most controls are designed to limit boost.

Blowoff Valve—The blowoff valve in Figure 10-2 is the simplest—and least-precise—method of sensing and controlling turbocharger speed. The valve is opened by turbine-inlet pressure only. Although this is related to compressor-discharge pressure and rotor speed, neither will be exactly controlled.

Another problem with this control is that regardless of the type of valve —poppet or flapper—it will flutter unless a damper is attached. If it flutters, the valve will destroy itself in a short time.

A blowoff valve can be mounted on the compressor-discharge duct, Figure 10-3. Again, this is not an exact control and is also subject to flutter.

This location is also only recom-

Hi-Tuned Datsun installation is an example of boost being controlled by carburetor size and exhaust restrictions.

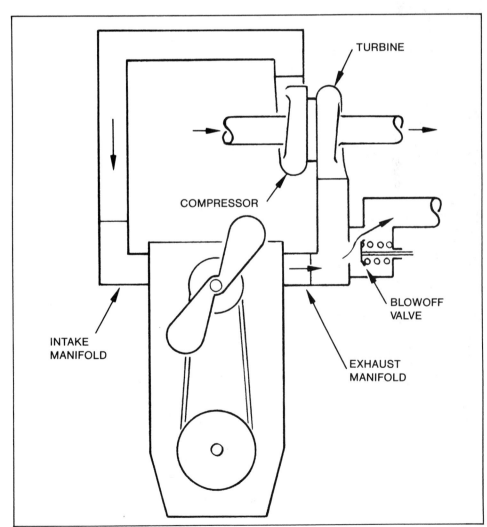

Figure 10-2—Turbocharged engine with simple exhaust-blowoff valve

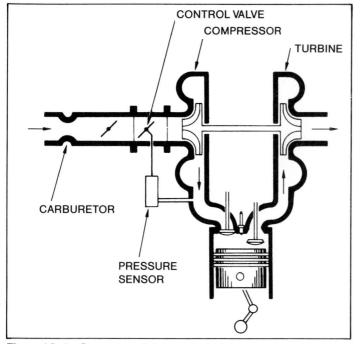

Figure 10-4—Compressor inlet-control system

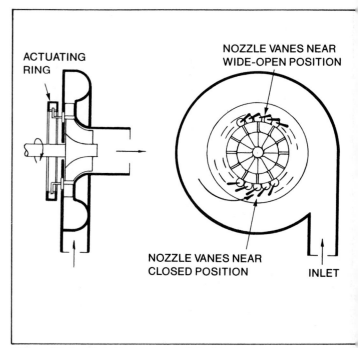

Figure 10-5—Variable-area-nozzle turbine

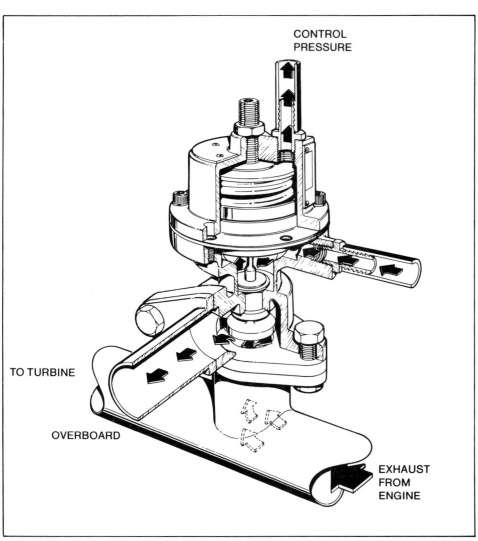

Rajay Wastegate is a simple and inexpensive poppet-type wastegate. Boost pressure is adjustable from 4 to 25 psi by changing springs. Valve can be made adjustable as shown in the drawing on page 85.

Crown blow-off valve mounts on intake manifold. Designed for use on fuel-injected engines, this type control should not be used on draw-through carbureted applications.

mended on a fuel-injected engine or on a blow-through application. These will release only air when the blowoff valve opens. With a draw-through installation, the blowoff valve will release fuel/air mixture into the engine compartment.

Compressor-discharge controls have been tried as a boost-limiting device by some racing officials, but racers are smart people. It didn't take them long to figure out that the valve could be fooled by using a larger compressor. When the flow of the compressor is much larger than the valve capacity, the valve no longer limits manifold pressure.

However, the officials got the last laugh. They required that the same compressor be used for the race as

Now out of production, Roto-Master BPR waste-gate cartridge in sandwich-type adapter fits turbine inlet of AiResearch and Roto-Master TO4 models and Rajay turbochargers.

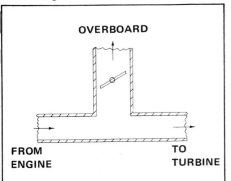

Figure 10-6—Butterfly-type waste gate

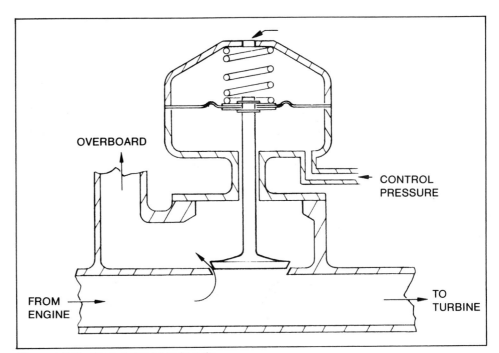

Figure 10-7—Poppet-type waste gate

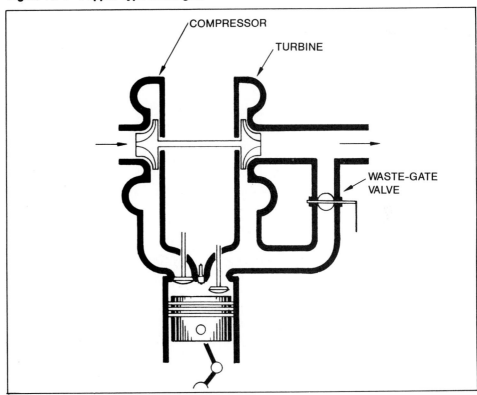

Figure 10-8—Manually operated waste gate

was used for qualifying. Excess fuel consumption was the result.

BAE uses a compressor blowoff valve on their Datsun 280Z and other fuel-injected engines, calling it an Absolute Pressure Control (APC).

Inlet Controls—A compressor-inlet-controlled system places a butterfly between the carburetor and the compressor, Figure 10-4. The valve reduces air flow to the compressor to limit manifold pressure.

An advantage of the inlet-controlled system is that the valve is on the cold side of the turbocharger. The major disadvantage is that manifold pressure is limited, but rotor speed may increase. It is not an effective way to control maximum turbocharger speed. When the pressure is being controlled, the higher compressor speed also raises intake-manifold temperature.

Variable-Area Nozzles—From a performance and efficiency standpoint, a variable-area nozzle, Figure 10-5, is

the best way to control turbocharger speed. With this method, all the gas goes through the turbine at all times and none is dumped overboard. The nozzle is opened and closed by an actuator controlled by the sensor.

This method is often used on air turbines. Unfortunately, it's expensive and the hot gases make it unreliable for turbochargers. The mechanism must be made from materials with

high strength and corrosion resistance at high temperatures. Even then, combustion products may jam the vanes.

Waste Gates—The most popular method of controlling turbocharger speed is with a waste gate or turbine-bypass valve. It can be either a butterfly, Figure 10-6, or poppet valve, Figure 10-7. These are covered in more detail later in this chapter.

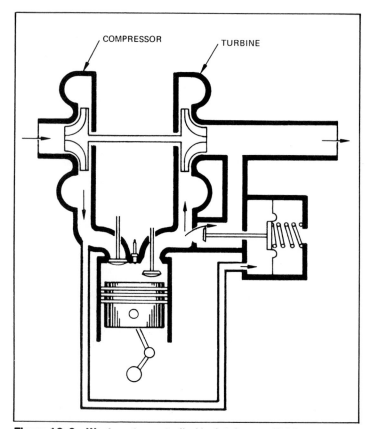

Figure 10-9—Waste gate controlled by intake-manifold pressure

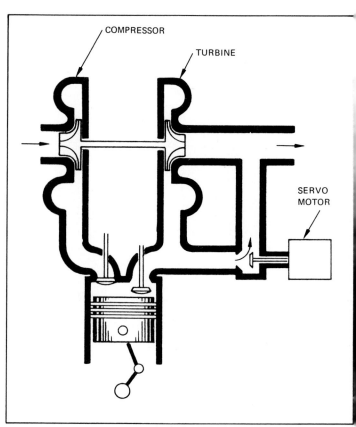

Figure 10-10—Waste gate controlled by servo motor

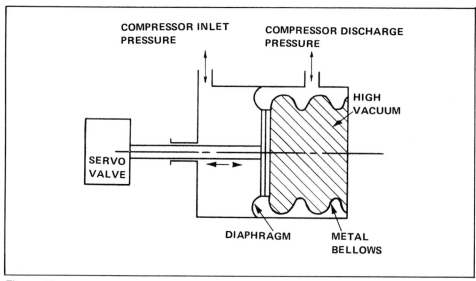

Figure 10-11—Pressure-ratio sensor

CONTROL SENSORS

The control valve can be actuated any number of ways. It may be operated manually, Figure 10-8; by intake-manifold pressure, Figure 10-9; or by a servo motor, Figure 10-10. The servo may be controlled manually or actuated by turbocharger speed, pressure ratio, gage pressure, absolute pressure, density or air flow.

Turbocharger speed can also be sensed electronically. By sending a sensor signal through amplifiers, the system can operate a waste gate to control turbocharger speed. The system requires an electric power source and rather elaborate electronics.

Pressure-Ratio Sensors—On a specific installation, pressure ratio is a direct function of speed, so turbocharger speed is normally sensed by a pressure-ratio sensor, Figure 10-11. This type of sensor is not open to the atmosphere, so it senses only the difference between inlet and outlet pressure.

If an engine with an uncontrolled turbocharger is moved from sea level to a higher altitude, turbocharger speed will increase. Pressure-ratio sensors were used extensively years ago, when turbochargers were run at the physical limits of the compressor impeller.

The pressure-ratio sensor prevents high-altitude speed increase. The engine can be operated at high altitudes without requiring any changes in the engine or turbocharger.

Absolute-Pressure Sensor—An absolute-pressure sensor, Figure 10-12, is similar to the differential-pressure sensor, but is independent of barometric pressure. The volume around the spring is enclosed in a bellows evacuated of air to create a high vacuum.

The vacuum around the spring not only gives an absolute-pressure reference but it is not affected by air-temperature changes. This sensor is

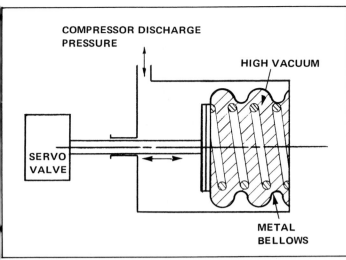

Figure 10-12—Absolute-pressure sensor

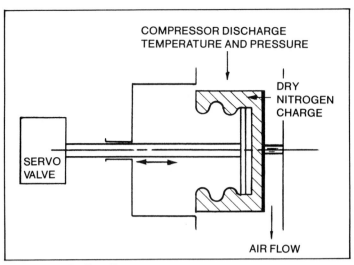

Figure 10-13—Density sensor

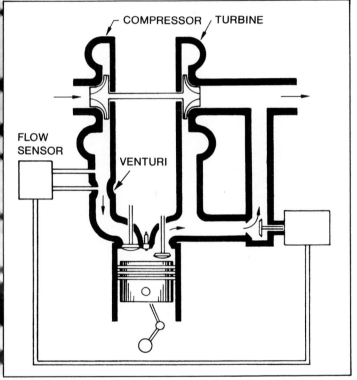

Figure 10-14—Waste gate controlled by flow sensor

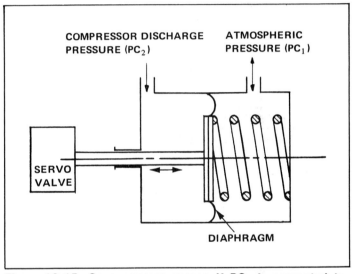

Figure 10-15—Gage-pressure sensor. If PC_1 is connected to compressor inlet, this becomes a differential-pressure sensor.

desirable when the engine is used at different altitudes, such as in an airplane. It has the big advantage of limiting intake-manifold pressure to the same absolute value regardless of altitude or barometric conditions.

Density Sensor—A density sensor, Figure 10-13, senses compressor-discharge temperature as well as pressure. It is placed in the discharge airstream of the compressor. These are a bit fancy for the type of controls we're talking about and normally would not be used in racing applications.

Flow-Control Sensor—A flow-control sensor, Figure 10-14, senses compressor air flow by the pressure drop across a venturi. This type of control allows having a higher boost pressure at low and medium speeds than at maximum engine speed. It is used where maximum torque is required at relatively low speed. This type of control gives a tremendous torque increase as the engine is slowed down.

Differential- or Gage-Pressure Sensor—A gage-pressure sensor, Figure 10-15, is the most popular sensor for engines operated mainly at

one altitude. When connected to a waste gate, this type sensor will start dumping exhaust gas when a preset intake-manifold gage pressure is reached. It will hold this pressure constant as long as the waste-gate capacity is adequate.

If PC_1 is connected to the compressor inlet, the gage shown in Figure 10-15 becomes a differential-pressure sensor. It will then sense the pressure differential, or difference, between the compressor inlet and outlet.

The IMPCO TC2 Turbocharger Pressure Control Valve, Figure

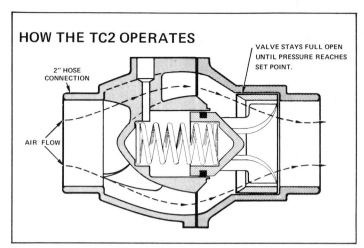

HOW THE TC2 OPERATES

2" HOSE CONNECTION

VALVE STAYS FULL OPEN UNTIL PRESSURE REACHES SET POINT.

AIR FLOW

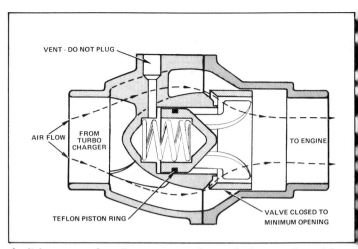

VENT - DO NOT PLUG

AIR FLOW

FROM TURBO CHARGER

TO ENGINE

TEFLON PISTON RING

VALVE CLOSED TO MINIMUM OPENING

Figure 10-16—TC2 control valve has spring-loaded piston/sleeve valve in an enclosed housing. Spring pressure keeps piston/sleeve valve open as shown. This allows full air flow from turbo at low boost.

As it increases, boost pressure works against cone-shaped face of piston. When pressure reaches predetermined control point, the spring compresses as shown. Piston/sleeve valve can move to close passage almost completely. This type of valve also limits rpm as it reduces volume of mixture supplied to engine.

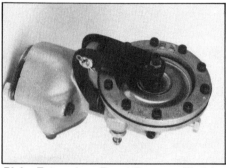

Blake Enterprises Boost Pressure Modulator (BPM) waste gate is used by several other kit manufacturers and installers. Valve is infinitely adjustable between 6- and 18-psi boost.

10-16, uses gage pressure to restrict air flow from the turbocharger. It is a rather simple boost-control device that can be placed between the compressor outlet and engine on an existing installation. This valve should be very reliable because it is on the cold side and has only two moving parts.

The IMPCO valve does, however, have a couple of disadvantages. Even when fully open, it restricts air flow to the engine. Consequently, it heats the mixture supplied to the engine. This increases the possibility of detonation. Also, if the compressor operates close to the surge line, the valve can cause compressor surge. This is a remote possibility, though, and would probably not occur on a passenger-car installation.

WASTE GATES

The combination of a gage-pressure sensor and a poppet-type waste gate is the most common method of control-

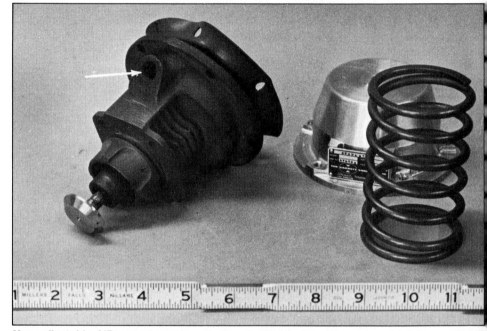

Non-adjustable AiResearch waste gate is actuated by boost pressure at intake manifold. Arrow indicates connection for control piping. Valve typically opens less than 1/8 inch to bypass exhaust and reduce turbine speed and boost pressure. Adjustable waste gates have a screw in the cover to adjust spring height and, thereby, boost pressure.

ling turbocharger speed. Where ambient conditions are fairly constant, such as on the street or a racetrack, this combination is simple and reliable.

If ambient conditions will change considerably, this combination has several disadvantages. For example, Ted Trevor has found that the altitude change during the Pike's Peak Hill Climb rules out using a gage-pressure sensor. Here he suggests an absolute-pressure sensor.

Another disadvantage with most waste gates controlled by manifold-

pressure is that they open the waste gate against the force of a spring, Figure 10-9. This requires a fairly large diaphragm to overcome the spring pressure. Also, unless the valve is damped, the spring and waste-gate valve can flutter, or chatter, at certain manifold pressures.

Servo-Operated Waste Gates—A more-exact method is to use intake-manifold pressure to operate a small hydraulic valve, Figure 10-17. The valve diverts oil pressure through the servo to open the waste gate.

This system has two advantages

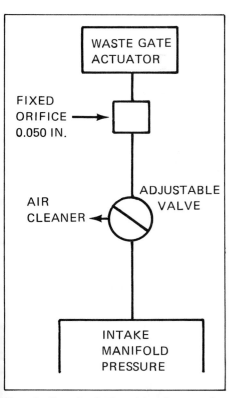

Here is the simplest and least-expensive way I know to make a waste gate adjustable. Valve can be a simple needle valve.

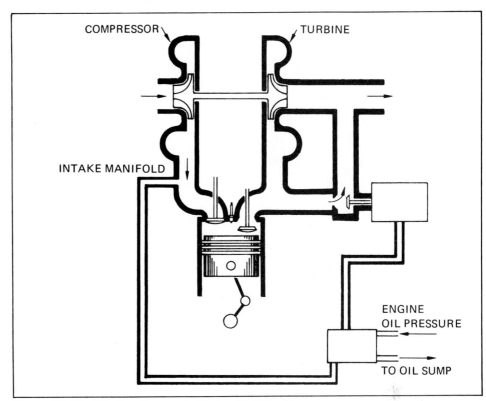

Figure 10-17—Waste gate operated by hydraulic servo

over the direct-operating waste gate. It can use a smaller diaphragm on the waste gate to control intake-manifold pressure. This is because the oil pressure is usually much higher than the intake-manifold pressure. Also, because the oil is incompressible, there is no possibility of the waste gate fluttering.

Unfortunately, the added complication of the hydro-pneumatic servo adds something else to go wrong, but engine damage can be prevented by spring-loading the waste gate. If the oil-pressure line fails, the valve moves to the full-open position.

Part-Throttle-Open Waste Gate—
For a waste gate to control boost pressure effectively, there must be some exhaust back pressure between the engine and the turbine. Under certain circumstances, back pressure on a turbocharged engine at ordinary road loads is often less than with a normal exhaust system. If an engine is equipped with a free-floating turbocharger that will not overboost the engine—even under maximum conditions—there will not be enough exhaust back pressure to make a waste gate effective.

A smaller turbine housing is always used with any waste-gate installation.

A smaller housing creates more back pressure on the engine, particularly at road load.

Unfortunately, the additional back pressure means an automobile equipped with a waste gate will ordinarily get worse fuel mileage than one without a waste gate. To prevent this loss in fuel mileage, it is best to have the waste gate open except when the turbocharger is needed for extreme acceleration or top speed.

Stanley Updike's part-throttle-open valve (PTO) waste gate, Patent 3257796, does this without complicated linkage, Figure 10-18. The right diaphragm is connected solidly to the valve stem while the left diaphragm is free to slide on the stem. A retaining ring on the valve stem allows the left diaphragm to open the valve but this diaphragm cannot close the valve.

The chamber above the right diaphragm is connected to the compressor inlet, so it senses the pressure drop through the carburetor. Chamber below left diaphragm is connected to the intake manifold and senses both vacuum and boost pressure.

At cruising power, the pressure drop through the carburetor creates a vacuum in the right chamber. When vacuum reaches 1.5 psig, it com-

presses the right spring and opens the valve.

During acceleration and at very low vehicle speeds, there is little or no drop through the carburetor. The right spring keeps the waste gate closed.

When the desired manifold pressure is reached, boost pressure underneath the left diaphragm compresses both springs. The waste gate opens to maintain a relatively constant intake-manifold pressure.

When connected to a small turbine housing, this PTO valve provides the same good low-end performance as a simple waste gate. Like a normal waste gate, it still prevents the engine from being overboosted at high engine speed.

Because the turbine is bypassed at normal cruising speed, the restriction in the exhaust is reduced. The engine "sees" very little back pressure and will get better fuel mileage. This improvement in fuel economy is gained at the expense of slower response time.

Another advantage of the PTO valve is that it opens and closes when the engine is accelerated. Every time the engine is started it goes from fully closed to fully open. An ordinary

Indy Car waste gate is mounted over transaxle and tucked in between engine and wing stanchion. Bolt in end of waste gate sets spring pressure. Lines from waste gate are routed to boost-control valve in cockpit. Photo by Tom Monroe.

Dual KKK waste gates are mounted on exhaust manifold with AiResearch TO4 turbo. Hoses from waste gates go to boost controller in cockpit. Photo by Tom Monroe.

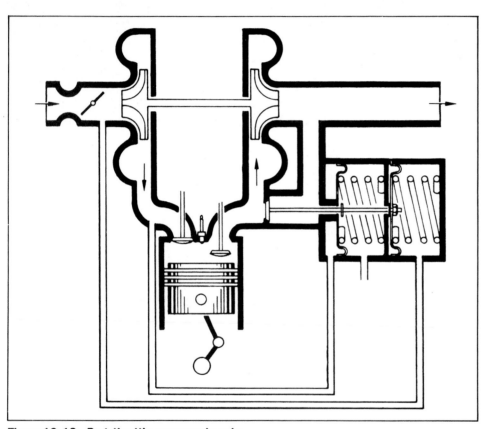

Figure 10-18—Part-throttle-open waste gate

waste gate may go for weeks without opening if the engine is never pushed. This inactivity is frequently the cause of waste-gate sticking.

Still another advantage of the PTO valve is that the turbocharger will run at a lower speed at cruise. The lower speed creates less pressure rise across the compressor. This allows the throttle to be opened farther. The larger throttle opening gives lower fuel/air-mixture temperatures, which give lower combustion temperatures.

Lower combustion temperatures reduce NO_x emissions and the possibility of detonation.

Carburetor Sizing—The use of a control not only improves the low-end performance of a turbocharged engine, but also makes the job of choosing a carburetor and tailpipe much easier. The carburetor can be much larger, because it will not be used as a restriction to limit boost.

A larger carburetor can produce other problems if not chosen carefully. Duke Hallock has found that when a large carburetor is used with a waste-gate-controlled turbocharger, compressor capacity should be 10 to 15% less than with a free-floating turbocharger.

One of the reasons for this is the smaller turbine nozzle A/R normally used with a waste gate. The higher turbine speed caused by the smaller turbine could drive a compressor into surge.

This problem will not occur with the free-floating system's larger housing, and would be remote in a light car. But it could definitely happen on a heavy vehicle climbing a long hill.

Turbine-Housing Sizing—According to Charles McInerney, a turbine housing that works best on a dynamometer will nearly always be about 10% too large for driving. This means about one A/R ratio.

For any given turbine, as A/R ratio decreases, turbine speed increases. This is true up to a point. If the A/R is too small, it will cause a large drop in the turbine efficiency.

The added back pressure on the engine may cause a power loss despite the additional intake-manifold pressure. If this should occur, use the next-smaller turbine wheel and go back to a larger A/R.

In general, a free-floating turbocharger/engine combination can give outstanding low-cost performance for the amount of work required. However, for ultimate performance—dragster, track racer, sports car, boat or airplane—some type of control is necessary. It's worth the added expense.

11 Intercooling

Whenever a group of engineers get together and someone brings up the subject of intercooling, invariably someone else says you mean *aftercooling*. Or if someone discusses aftercooling, another person says *intercooling*. Regardless, they are usually referring to cooling the air or fuel/air mixture somewhere between the compressor discharge and the engine. Many engineers use the all-inclusive term, *charge-air cooling*.

INTERCOOLING ADVANTAGES

As noted in the chapter on turbocharger design, any increase in air pressure also brings a corresponding increase in temperature. With a turbocharger, the amount of the increase depends a great deal on the amount of boost and the compressor efficiency.

Increased temperature cancels out some of the benefits of the higher pressure because the heated charge is less dense. Cooling the charge regains some of the density and brings other benefits as well.

If we had a perfect heat exchanger, charge-air temperature could be reduced to that of the cooling medium without pressure drop. This is not possible because there will always be a pressure drop through the heat exchanger. It is also not possible to lower the charge temperature to the cooling-medium temperature.

The charge is cooled by ducting it to a heat exchanger. The cooling medium may be ambient air, Freon, ice water, sea water, engine coolant, or almost any medium that is consistently cooler than the charge, Figure 11-1.

The temperature drop of the charge air passing through the heat exchanger is a function of several factors. It varies with heat-exchanger size and the temperature and available flow rate of the cooling medium.

There is always a pressure drop of the charge as it goes through the heat exchanger. The amount of pressure drop must be weighed against the temperature drop. For example, there is no advantage to lowering the charge temperature 100F and then losing

Intercoolers displace radiator at front of GTX Mustang. Radiators—two are used—are installed in rear quarters. This illustrates importance of intercooling with turbocharged engines. Zakspeed-modified 1700cc Kent-based four-cylinder Ford produces about 550 HP at 60-in.Hg-maximum boost. Photo by Tom Monroe.

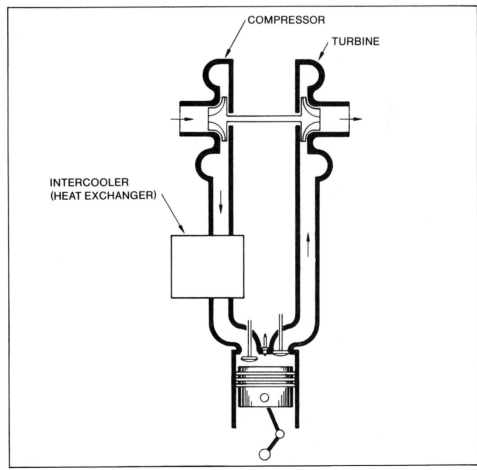

Figure 11-1—Turbocharged engine with intercooler

Flat-6 2.4-liter Porsche is turbocharged with two Roto-Master turbos. Andy Johnston fabricated an air-to-air intercooler that incorporates a Mack tip-turbine fan to circulate cooling air through heat exchanger.

Renault R5 turbo built by Katech, Inc. Engine uses two-stage air-to-air and liquid-to-air intercooler that was originally designed by Renault for their Formula-1 car. See drawing at right.

half the pressure. If the charge air passing through the heat exchanger does not become considerably more dense, the heat exchanger is doing only half its job.

Reducing Exhaust Temperature— As a rule of thumb, one-degree decrease in intake-manifold temperature will give a one-degree drop in exhaust temperature. For example, if an intercooler reduces an engine's intake-manifold temperature by 100F, the exhaust temperature will go from, say, 1500F to 1400F.

This heat reduction helps the exhaust valves. It also cuts down on the engine's heat-rejection requirement, unless jacket water is used as the heat-exchanger's cooling medium.

Pressure Versus Density—Higher charge density will allow more *mass* of air per minute to flow through the engine at any given intake-manifold pressure. This means more fuel can be burned and the engine can produce more horsepower.

As an example, I'll use a 151-CID engine running at 10,000 rpm. At an intake-manifold pressure of 45 psig, the engine produces 900 HP. Assuming 65% compressor efficiency, the intake-manifold temperature will be 487F. Mass air flow through the engine will be about 900 pounds per hour.

If a heat exchanger with only 50% effectiveness is placed between the compressor discharge and the engine, on a 100F day the heat exchanger will reduce intake-manifold temperature to 293F.

Even if the intercooler lowers intake-manifold pressure to 35 psig, mass flow will increase to 941 pounds per hour. In addition, for the same horsepower and fuel/air ratio, exhaust temperature will be reduced approximately 194F.

There are two advantages to the engine. First, the engine's overall operating temperature will be reduced. Second, the combustion-chamber pressure for a given *brake mean effective pressure* (bmep), or working pressure, will also be reduced. These combine to lessen stress on the engine.

Gasoline engines are sensitive to charge temperature because of preignition and detonation limits. With an intercooler, there is less tendency toward detonation, so the octane requirement for a given engine output can be reduced.

Intercooling also reduces the engine's tendency to develop "hot spots" in the combustion chamber. This reduces the possibility of preignition.

Computing Heat-Exchanger Effec- tiveness—As another example, the charge-air temperature is 200F and the cooling-medium temperature is 100F. If we are able to lower the charge temperature to 150F, the heat exchanger has an effectiveness of 50%.

$$\frac{200F - 150F}{200F - 100F} = \frac{50}{100} = 0.5 \text{ or } 50\%$$

If a heat exchanger is used that can drop the charge temperature to 130F, the effectiveness is 70%.

$$\frac{200F - 130F}{200F - 100F} = \frac{70}{100} = 0.7 \text{ or } 70\%$$

A value of 70% is reasonable and is used for examples in this chapter.

Cooling Media—Assuming a compressor-discharge temperature of 250F and an ambient temperature of 75F, a 70% effective air-to-air heat exchanger will reduce the charge air temperature (250 − 75) x 0.7 = 122.5F. Charge temperature leaving the intercooler will be 250 − 122.5 = 127.5F.

If jacket water is used as the cooling media, at 180F the temperature drop will be (250 − 180) x 0.7 = 49F. This results in a charge-air temperature of 201F. This small decrease makes it impractical to use jacket water unless the compressor-discharge temperature is at least 300F.

Using a low-temperature liquid,

such as sea water at 75F, will result in the same 127.5F charge-air temperature as an air-to-air heat exchanger. The advantage is that the air-to-liquid type is much smaller and does not require a fan to circulate the cooling medium.

Hans Mezger of Porsche has made good use of air-to-air heat exchangers on several Porsche racing and passenger cars. This system was used on the type 935 and 936 racing cars. One system used water in the intercooler and the other used air.

Hans is probably the best example of an engineer who did not let all that talk about turbo lag and why turbocharged engines were no good in road racing deter him. Porsche's turbocharged and intercooled cars have dominated everything they have entered. Finally, the rest of the pack has decided that maybe intercooling isn't a bad idea.

In 1980, Porsche built and tested a turbocharged/intercooled car for the Indianapolis 500. USAC officials observed the car's performance and immediately changed their rules to reduce the allowable boost pressure. The car was withdrawn from the race.

And then there is the intercooled turbo-Renault 1983 Formula-1 car. It produced approximately 800 HP from a 90-CID engine! Enough said.

A drag racer or a Bonneville-type car can carry enough low-temperature liquid to cool the charge, in much the same way that a tank of liquid cools the engine. The tank could contain ice water, acetone or alcohol with chunks of dry ice. Be careful with cooling media below 32F because the charge side might "ice up" on a humid day. Tom Keosabababian successfully used a Freon spray on one of his intercooler installations.

Computing Charge-Air Temperature—Let's start with a 300-CID engine turbocharged to 15 psig at sea level, with a 15 psia barometer and 80F ambient temperature. The engine is run on a dynamometer and produces 350 HP at some speed. Assuming a compressor efficiency of 70%, you can calculate the compressor-discharge temperature.

The compressor pressure ratio is:

$$\frac{\text{Compressor-Outlet Absolute Pressure}}{\text{Compressor-Inlet Absolute Pressure}} =$$

$$\frac{15 + 15}{15} = \frac{2}{1}$$

Refer to Table 1, page 142.

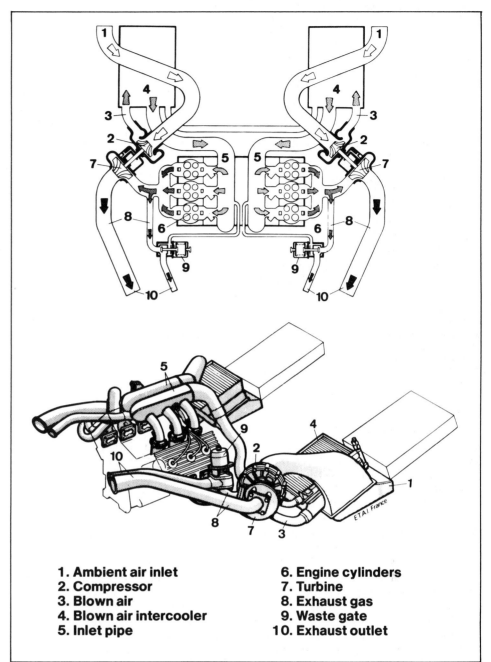

1. Ambient air inlet
2. Compressor
3. Blown air
4. Blown air intercooler
5. Inlet pipe
6. Engine cylinders
7. Turbine
8. Exhaust gas
9. Waste gate
10. Exhaust outlet

Dual-turbo intercooled Formula-1 Renault. Drawing courtesy of Renault.

When
r = 2
Y = 0.217

Ideal Temp. Rise

= (80 + 460) x 0.217
= 540 x 0.217 = 117F

Actual Temp. Rise

$$= \frac{\text{Ideal Temp. Rise}}{\text{Compressor Efficiency}}$$

$$= \frac{117F}{0.7} = 167F$$

Compressor-Discharge Temp.

= 80F + 167F
= 247F

If you place a 70% effective air-to-air heat exchanger between the compressor discharge and the intake manifold, it will lower the charge temperature

(247F - 80F) x 0.7 = 167 x 0.7 = 117F.

The temperature entering the engine will now be 247F − 117F = 130F.

Because the heat exchanger is not ideal, there will be a loss of pressure through it. To be conservative, assume a 1-psi loss. Using these figures, the density increase with the heat exchanger is:

$$\frac{247 + 460}{130 + 460} \times \frac{15 + 14}{15 + 15} = \frac{707}{590} \times \frac{29}{30} = 1.158.$$

As a result, the intercooled engine

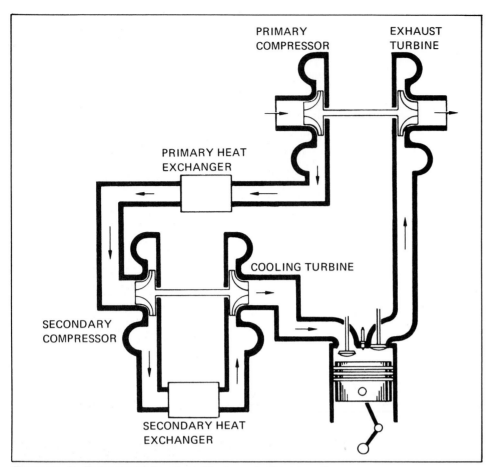

Figure 11-2—Turbocharged engine with air-cycle charge cooling

Two Mack Truck intercooler cores were used by Kinsler Fuel Injection for this installation. Cores are placed in series; intake air passes through one, then the other. Slightly higher pressure drop results with this arrangement—as opposed to a parallel setup. Increased pressure drop is more than offset by a much larger temperature drop.

can put out 18% more power at the same speed with 1 psi less manifold pressure. However, there is a catch to this.

Intercoolers & Turbine Speed—In some cases, the exhaust temperature and pressure will drop in proportion to the intake-manifold conditions. The turbocharger then slows, further reducing intake-manifold pressure. When this occurs, you must use a smaller turbine housing to maintain the same boost pressure.

Multiple Installations—You can cool the charge below the temperature of the cooling medium. If the gas is compressed further with a secondary compressor, it can then be cooled and expanded through a turbine used to drive the secondary compressor, Figure 11-2. This method is used successfully on some large stationary diesels, but is complicated for a small engine.

Jim Kinsler has been working with intercoolers for more than 15 years and has done a number of multiple installations. According to Kinsler, in certain applications, parallel-core installations (having the charge split through two separate intercoolers) can lower the mixture velocity so much that poor intercooler effectiveness results. Kinsler always uses series (one cooler feeding into another) installations for any engine displacing less than 400 cubic inches.

It is possible to connect turbochargers in series and obtain intake-manifold pressures over 200 psig by

Ford Motorsports' intercooler kit for Turbo Coupe T-Bird: Intercooler increases engine power by supplying denser air charge to engine and prolongs engine life by reducing or eliminating detonation by reducing intake air-charge temperature. On production cars with detonation sensors that automatically retard ignition, a noticeable power increase can actually be felt as spark retard is reduced or eliminated. Photo by Tom Monroe.

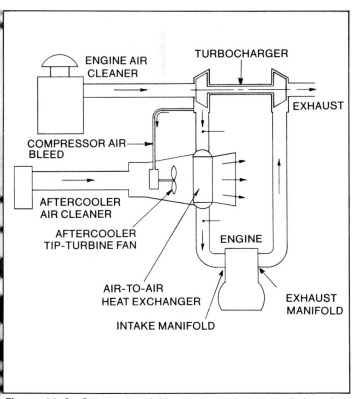

Figure 11-3—Schematic of Mack air-to-air intercooled engine with tip-turbine fan

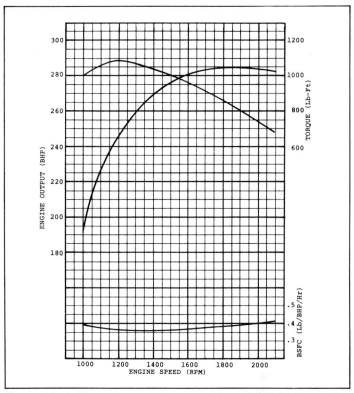

Figure 11-4—Horsepower and torque—Mack ENDT 676 with air-to-air intercooler

using heat exchangers between the compressors and between the final compressor and the engine. This method is discussed in detail in Chapter 16, Tractor Pulling.

WHERE TO FIND THEM

Things aren't as bad as they were a few years ago when it comes to buying a heat exchanger. Gale Banks and Tom Scahill make bolt-on liquid-to-air heat exchangers for popular automotive engines. These are designed for marine use, but they will work well on a drag racer or Bonneville car. They can be adapted to use ice water circulated with a pump.

There is an air-to-air heat exchanger available for those willing to pay the price. Mack Truck uses an air-to-air heat exchanger on a version of their 672-CID diesel engine, Figure 11-3. This engine produces 285 HP at 2100 rpm and has a 55% torque rise to 1200 rpm.

Cooling air is forced through the aftercooler by a fan with an axial turbine attached to its periphery, Figure 11-4. The turbine is driven by a small amount of air bled off the turbocharger compressor.

Jim Kinsler tested the Mack Truck air-to-air cores against 14 other existing brands and types. He found that the Mack Truck cores were significantly better than any others, both in terms of high heat transfer and low pressure restriction. Kinsler has used the Mack cores in a number of automotive racing installations. He markets these and a new line of cores from Blackstone that are also top performers.

Kinsler warns to be careful to keep the charge-air velocity through the

Gale Banks adapted his Venturi-Flow intercooler to the dual-turboed big-block Chevrolet-powered Kehoe-McKinney-Banks Bonneville car. Car set an average two-way record of 240 mph at the Bonneville Nationals.

cores fairly high. High mixture velocity promotes good "scrubbing" of the cooling surfaces, which makes the unit more effective.

Reeves Callaway and HKS offer bolt-on kits for factory turbo cars. They also sell components for building your own intercooler system.

DAECO Intercooling System— Alvin Lowi, Technical Director of

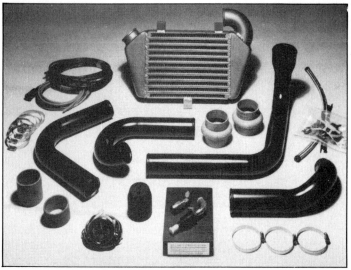

Janspeed intercooler uses air-conditioning system to cool intake charge on this Jaguar. Valves control Freon flow from air-conditioning system through an evaporator core inside intercooler.

Callaway Turbosystems kit includes ducting, fuel-enrichment system, boost gage and hardware. Photo by C. E. Green.

Daigh Automotive Engineering Corporation (DAECO), has gone one or two steps forward with intercooling. Lowi has combined engine and charge cooling into one system. It may even be possible to add the air-conditioning system, but that's another step.

Figure 11-5—Schematic of DAECO two-phase engine- and charge-cooling system

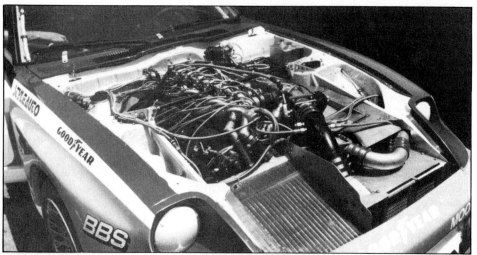

Front of Electromotive Datsun 280ZX IMSA GT car shares radiator (left) and intercooler (right).

Figure 11-5 is a schematic of the DAECO (patents pending) system. To oversimplify the system, it is similar to the old gas-fired refrigerators. A refrigerant such as Freon is used to cool the engine.

The refrigerant boils in the cooling jacket. As it leaves the engine, the high-pressure vapor is fed through a jet compressor. The compressor is little more than a venturi, which creates a vacuum as the high-pressure vapor passes through it.

The velocity of the jet aspirates, or draws out, the low-pressure vapor after it has cooled the charge. The jet compresses the vapor up to 5 psig so it will condense when it is cooled in the radiator. The resulting liquid is collected in a receiver tank.

Some of this liquid then passes through an expansion valve and back to the intercooler. The rest is pumped back into the engine cooling jacket.

The big advantage of this system is that the cooling medium experiences phase changes, much like a conventional air-conditioning system. As a chemical changes from one phase to another, e.g., liquid to gas, there is a large temperature change. Consequently, the system can operate at widely different temperatures, depending on whether it is used to cool the engine or the charge.

SUMMARY

The advantages of charge cooling include lower temperatures, lower pressures and higher potential output. The combination makes it very attractive, particularly at pressure ratios of 2:1 and higher.

12 Marine Engines

Ocean-racing boat at full song. Engine compartment is shown on page 95.

In some ways, marine engines are a lot easier to turbocharge than automobile engines. In other ways, they pose problems not encountered in automobiles.

In an automobile, engine speed at full throttle is limited by the instantaneous automobile speed and the gear ratio. This is true even with engines driving through a torque converter.

This is not true with a marine engine. An engine mounted in a propeller-driven boat can be accelerated to near maximum speed long before the boat itself reaches its maximum speed. In a jet boat, the engine can be revved to peak speed while the boat is tied to the dock.

MARINE REQUIREMENTS

On a boat, power can be fully utilized only when it coincides with the

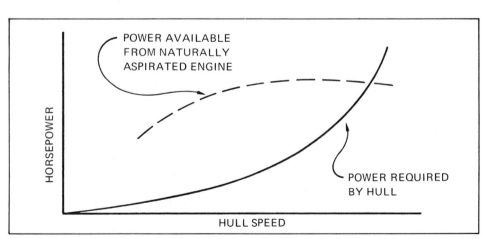

Figure 12-1—Engine power available compared to hull requirement with a naturally aspirated engine

power required by the hull. Figure 12-1 illustrates this for a non-planing hull. Although every hull shape has a

different curve, the situation is about the same.

Except in the case of a drag boat, if

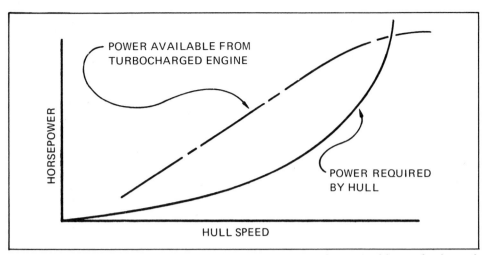

Figure 12-2—Engine power available compared to hull requirement with a turbocharged engine

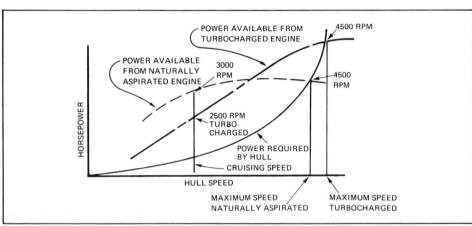

Figure 12-3—Naturally aspirated and turbocharged engines matched to same hull with same maximum engine speed

the engine can approach full speed regardless of hull speed, the extra power available at low hull speeds is of little use.

For example, suppose a boat is cruising at 20 knots and 2200 rpm and the throttle is opened fully. If the engine immediately accelerates to 4400 rpm and remains there when the boat reaches 30 knots, then the extra torque available at 2200 rpm is wasted.

Turbocharging a marine engine makes it possible to match the power curve more closely to hull requirement, Figure 12-2. This results in a more efficient system.

In Figure 12-3, available-power curves are plotted for both naturally aspirated and turbocharged engines in the same hull. Propellers are matched to the engines to make both power curves cross the hull-requirement curve at 4500 rpm. This is done by using a higher-pitch propeller on the turbocharged engine.

If the boat slows to cruising speed, note that the turbocharged engine will

be running 500 rpm slower than the naturally aspirated engine. This gives better fuel consumption and longer engine life.

Also note that the turbocharged engine will drive the hull considerably faster than the naturally aspirated engine. The higher output of the turbocharged engine allows using a higher-pitch propeller. Consequently, the higher boat speed can be reached at a lower rpm.

This low-rpm power advantage has been demonstrated many times in long ocean races. Turbocharged engines running approximately 4500 rpm had the same power output as naturally aspirated engines at 7000 rpm. The rpm difference meant the turbocharged engines were able to outlast the naturally aspirated engines.

A few years ago, turbocharged Daytona Engines so dominated ocean racing that turbochargers were banned. In 1978, the Pacific Offshore Powerboat Racing Association reinstated turbochargers in the open

classes with no displacement penalty.

Because of the difference in torque requirements, matching a turbocharger to a marine engine is somewhat different than on a automobile engine. Instead of a broad torque curve, a marine engine should give a linear power increase with increased rpm.

The turbocharger and turbine housing are sized to obtain maximum compressor efficiency at maximum engine speed. With the same engine size and maximum speed, the marine engine will normally use a larger compressor than the automobile engine.

In the case of the automobile engine, the engine's mid-range corresponds to the area of maximum compressor efficiency, Figure 12-4. This allows the engine to produce higher torque at mid-range. The high-speed end of the power line extends into the lower-efficiency region of the compressor.

On the marine engine, Figure 12-5, a larger compressor is used because there is rarely any high-torque requirement at mid-range. The high-speed end goes through the center of the highest efficiency island.

Special Hull Requirements—Normally, supercharging is only required at high speed. One exception is when the hull requirement has a hump in it, Figure 12-6. Here, the turbocharger must be sized to allow the engine to produce enough power to accelerate the hull through the hump. See graph on following page.

Curve 1 is an example of a turbocharged engine mismatched to this hull. The engine has plenty of power at the top end, but the power will never be used. The shaded area shows where hull requirement is higher than power output. The hull will never reach planing speed with this engine.

Curve 2 shows the turbocharger matched to give more low-end power to get the hull over the hump. If maximum speed is most important with this type of hull, it may be necessary to use a waste gate and smaller turbine housing. This combination will allow more boost at low speed, without overboosting at maximum speed.

Tom Scahill solved this problem in a different manner. He adapted an automatic transmission to the engine and Vee drive. The torque converter has been replaced with a solid coupling and he only uses two forward speeds and reverse. Tom says, "A lot of

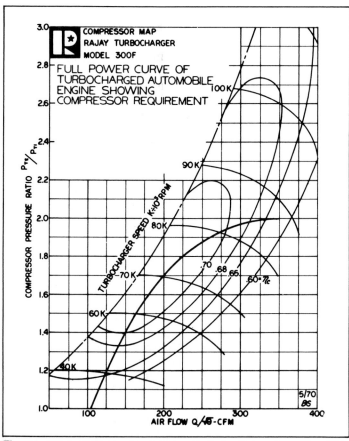

Figure 12-4—Full-power curve of turbocharged automobile engine showing compressor requirement

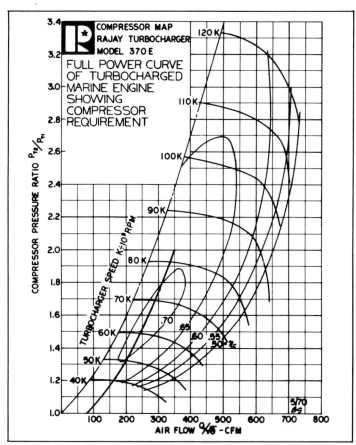

Figure 12-5—Full-power curve of turbocharged marine engine showing compressor requirement

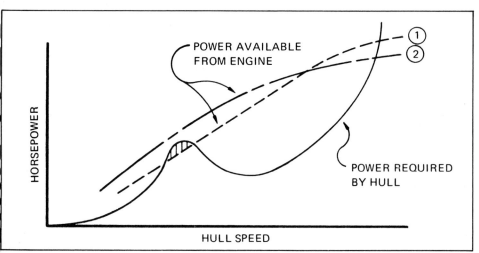

Figure 12-6—Turbocharged engine incorrectly (1) and correctly (2) matched to hull with special power requirements

Two Race Aero twin-turbocharged 454 Chevrolets combine to produce over 1600 HP. Engines are staggered to keep drive shafts close together.

heads turn when they hear me shift!''

Water-Cooled Exhaust—On a racing boat with dry exhaust manifolds, the correct turbine-housing size can be predicted without difficulty. The exhaust energy available will be similar to an automobile engine.

This is not true of a pleasure boat with a water-cooled exhaust system. To begin with, the water jacket around the exhaust manifold will reduce the exhaust-gas temperature at least 200F. The cooler exhaust reduces the volume flow available to the turbine.

Water-cooled exhaust manifolds usually have a water trap after the turbine. This prevents sea water from backing into the turbocharger when the engine is not running. The trap creates back pressure on the turbine. In addition, engine-cooling water dumped into the exhaust pipe after the trap adds still more back pressure to the turbine.

Jim Steen used a GM/Harrison air-conditioning evaporater core to intercool his dual-turbocharged small-block Chevrolet ski boat. Hobbs switch turns on water injection at high boost. Photo by Tom Monroe.

Tom Scahill turbocharged this 427-CID Chevrolet-powered 20-foot Rayson Craft using two Rajay units and his own intercooler. Top speed is over 110 mph.

Banks Stage V offshore-racing 454-CID big-block Chevy produces 915 BHP on aviation gas. Dual turbos draw through two Holley four-barrels, then blow through a dual-element intercooler.

Stage III Banks 454-CID Chevy is used for military-, commercial- and pleasure-craft applications. Engine produces 540 BHP when set up for running on 88-octane gas; 640 BHP with 91 octane.

To compensate for the back pressure, a smaller turbine housing is used. Size, of course, also depends on the conditions encountered in the particular installation.

MARINE INTERCOOLING

Because of the availability of cool water, intercooling is practical on marine engines. With 70F to 80F water as a cooling medium, an extremely effective intercooler can be added easily. Marine intercoolers can be small enough not to increase the overall size envelope of the engine.

Intercooling is discussed in detail in Chapter 11, but has several advantages in marine use.

Intercooling reduces the octane requirement of a turbocharged engine. In some cases, intercooling has allowed engines to be set up to run turbocharged on regular-grade gasoline.

Intercoolers are now virtually considered necessary to get the most out of a turbocharged engine. However, they are not without problems of their own, particularly on a marine engine. The problem is fuel condensation.

Carburetor Heating—One of the larg-est uses of turbocharged marine engines has been in sport-fishing boats. Most owners of these boats like to get out to the fishing grounds fast, do their fishing at leisure, and get back home fast. Out and back are no problem, but leisurely fishing gives the trouble.

Suppose the engine is idled to trolling speed for a half hour. When idling, manifold vacuum is very high and no supercharging takes place.

Under these conditions, the fuel/air mixture is already quite cool when it reaches the intercooler. But

Sabre Marine two-stage turbocharging with intercooling and aftercooling produces 450 HP from 6-liter (366 cubic-inch) diesels.

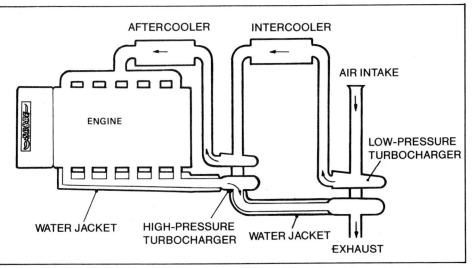

Figure 12-7—Schematic of turbochargers and intercoolers in the Sabre 420 Diesel. Note similarity to staged tractor-pull engine, Figure 16-2.

M & W kit with water-jacketed turbine housing on MerCruiser (Chevrolet) six

ambient.

Water-heated carburetor adapters are available from Ak Miller, Tom Scahill, Turbo City, Spearco and Gale Banks Engineering.

Marine Diesels—Not long ago, diesel engines in ocean racing were a rarity. Once in a while a large distributor of diesel engines would enter one in a long-distance race, but it was more for publicity than the chance of winning.

The main problem was specific output—weight compared to output. A very good diesel engine will weigh 5 lb/HP. This makes it rough to compete against gasoline engines, which weigh as little as 2 lb/HP.

Sabre Engines LTD has been making high-output marine diesel engines for years but could not come near the 2 lb/HP figure—until recently. Perhaps their engineers attended a tractor pull (see Chapter 16) or were just willing to try the ultimate. Either way, the results were impressive. The installation is similar to that on some pulling tractors, even though the application is quite different.

There are two major differences between this system, Figure 12-7, and one used on a pulling tractor. First, the Sabre diesel must run for hours, not just 30 seconds or so. Second, there is an unlimited heat sink for the intercooler and aftercooler. The pulling tractors using this system must rely on a fixed amount of ice and water.

At full power, intake-manifold pressure is over 50 psig. Obtaining pressure this high in a single stage with reasonable efficiency and range is difficult. Staging gives both high efficiency and broad operating range.

the intercooler cools it still further with 70F or 80F water. This and the low mixture velocity causes fuel condensation in the intake manifold. After a while, a puddle of fuel collects.

All of a sudden, one of the fishermen gets a strike and hollers for full power. When the throttles are opened, the engine does not respond instantly. This can be very annoying, particularly if you lose a big fish.

Ted Naftzger had this problem and solved it by making two changes. First, he added a water jacket to the bottom of the carburetor box. This allows engine-jacket water to warm the carburetor—much the same as on an automobile engine.

Naftzger then added a bypass valve around the intercooler. This allows him to shut off the cold water and prevent fuel from condensing in the intake manifold.

Ted used a manual bypass valve, but a diaphragm-operated valve could be used. The valve could allow water to flow through the intercooler only when manifold pressure was above

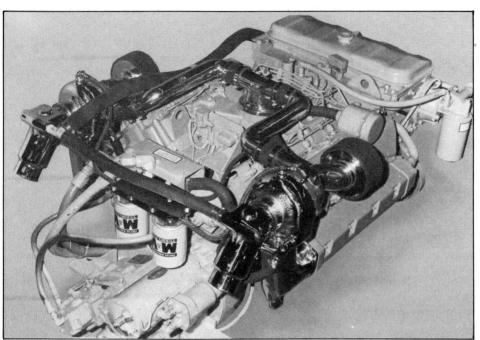

M & W turbocharged Caterpillar 3208 diesel performs like a naturally aspirated gasoline engine of the same displacement.

Three-stage turbocharging on Allison V1710 engine in an unlimited hydroplane. Output is estimated at 3500 HP. Photo by Jim Dunn.

With staged turbochargers, Sabre was able to run the turbochargers at a little over 2:1 pressure ratio in each stage. Under full power, intake-manifold temperature is maintained at 75F with sea water at 54F.

Two of these engines were installed in Romans Sabre, an aluminum catamaran. The boat won the 1980 European Class II Championship and the British Championship for both Class I and II.

Marine Turbocharger Kits—M & W Gear Company has been making bolt-on turbocharger kits for farm tractors since 1962. There's no doubt they make good kits—they've sold more than 100,000.

In 1974, M & W entered the marine field with bolt-on kits for 4- and 6-cylinder Chevrolet engines. These engines are supplied by both MerCruiser and Chris Craft. Engineer Jack Bradford was as thorough on these as he was on M & W's high-quality diesel kits.

These kits do not contain intercoolers, so they are for a 20% to 25% increase in power and are not designed for racing. They are, however, honest-to-goodness do-it-yourself bolt-on units.

Currently, M & W Gear Company makes marine-engine turbocharger kits for all MerCruiser Chevrolet and Ford applications, OMC 6-cylinder (Chevrolet), small- and big-block Chrysler, 460 Ford (as supplied by Indmar, Hardin, Holman-Moody) and the Caterpillar 3208 diesel.

Gale Banks Engineering has been making custom turbocharger installations on marine engines since 1970. This entire business has developed around turbocharging. Gale Banks is now the foremost supplier of high-performance turbocharger kits and complete turbocharged engines for marine applications—both pleasure and racing.

Kits and complete turbocharged engines include: Chevrolet big-block 396, 427 and 454; small-block Chevrolet; Ford big-block 429 and 460; Oldsmobile 403 and 455. Additionally, kits are available for the entire MerCruiser V8 engine line.

Banks offers numerous accessories for the marine enthusiast, including watercooled exhaust manifolds, intercoolers, watercooled turbo housings and shields, oil pans with turbo drains, oil pumps and oil coolers.

Turbocharged marine engines are now widely used in closed-course, drag and offshore racing. Turbochargers have almost completely replaced mechanical superchargers on unlimited hydroplanes. Typical output of 3500 HP at 4000 rpm is seen on the Allison V1710 engines.

13 Two-Stroke Engines

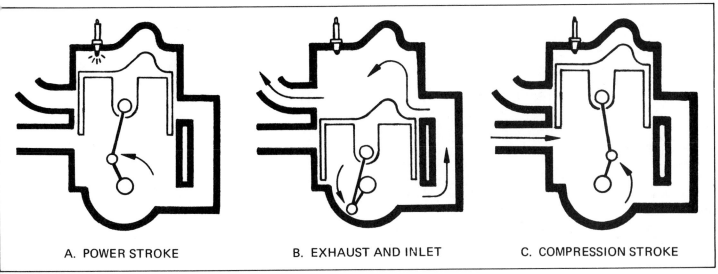

A. POWER STROKE   B. EXHAUST AND INLET   C. COMPRESSION STROKE

Figure 13-1—Simple two-stroke engine

When discussing the hows and whys of turbocharging, we normally think of the standard four-stroke engine—passenger-car, racing, marine or aircraft. But two-stroke engines can't be ignored because they are used in motorcycles, snowmobiles, all-terrain vehicles, boats and recently even in airplanes.

Two-stroke engines can and have been successfully turbocharged, but the conditions are different than on a four-stroke engine. Therefore, many of the problems have to be attacked differently.

TWO-STROKE BASICS

Except for the Wankel engine, four-stroke engines can be considered basically the same as to valve location and combustion-chamber shape. Not so with two-stroke engines.

Types of Two-Stroke Engines—Two-stroke engines are manufactured with many different types of porting and scavenging. Each must be treated differently when it comes to turbocharging.

The simplest of the two-stroke engines is often referred to as a *cross-flow crankcase-scavenged* type, Figure 13-1. The fresh charge from the crankcase enters one side of the cylinder while the exhaust leaves the other.

When the engine starts its power stroke, both the inlet and exhaust

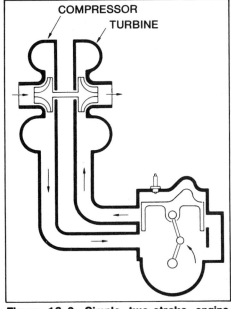

COMPRESSOR   TURBINE

Figure 13-2—Simple two-stroke engine with turbocharger

ports are covered by the piston, Figure 13-1A. The crankcase port is open. Once the engine completes its power stroke, the exhaust port, which is slightly higher than the inlet port, allows cylinder pressure to drop to atmospheric, Figure 13-1B.

When the inlet port is uncovered, the pressure in the sealed crankcase forces the fuel/air mixture into the cylinder. The crankcase port is covered to prevent the charge from escap-

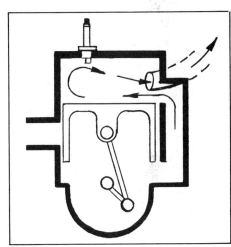

Figure 13-3—Loop-scavenged two-stroke engine

ing back to the carburetor.

The baffle on the top of the piston directs the fresh charge up across the top of the combustion chamber. This prevents most of the charge from escaping through the exhaust port.

When the piston is near the end of its compression stroke, the crankcase port is uncovered, Figure 13-1C. The partial vacuum in the crankcase draws fuel/air mixture into the crankcase through the crankcase port.

Many variations of this design include reed or flapper-type check valves on the crankcase instead of the crankcase port. Another variation

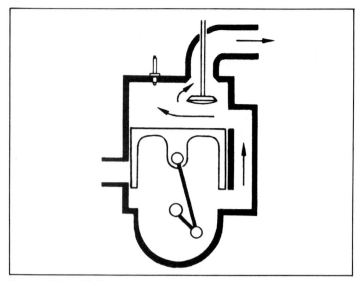

Figure 13-4—Uniflow-scavenged two-stroke engine

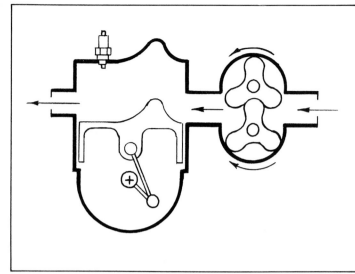

Figure 13-5—Two-stroke engine with engine-driven scavenge pump

mates the crankcase port to a rotary valve in or on the crankcase.

None of these variations has much effect on the problems associated with turbocharging this type of engine, Figure 13-2.

The main problem is that the exhaust port opens before the inlet port and closes after it. This means supercharge pressure is determined by the back pressure from the turbine. As a result, it is possible for some burned exhaust gases to re-enter the cylinder after the inlet port has closed.

Both problems can have a substantial effect on the volumetric efficiency of the engine. Actual results are often different than those calculated.

Another type of two-stroke engine is the *loop-scavenged* type, Figure 13-3. Intake and exhaust ports are not opposite one another, so no baffle is required on the top of the piston. It still has the problem of the exhaust port closing after the inlet port.

On a *uniflow-scavenged* two-stroke engine, the exhaust port is closed by a poppet valve such as used in four-stroke engines, Figure 13-4. This valve is opened and closed by a camshaft and can be timed to open and close before the intake port opens. The valve prevents the fuel/air mixture from being blown out the exhaust port or exhaust gas from re-entering the cylinder when it is not wanted.

Many other types of two-stroke engines include opposed-piston engines, but the three described here cover most cases.

It is not necessary to use the crankcase for a scavenge pump, but it is convenient and inexpensive on small engines. Using the crankcase for this purpose means a separate crankcase must be provided for each cylinder, except where two cylinders fire at the same time.

The efficiency of the crankcase as a scavenge pump is inversely proportional to the volume of the crankcase when the piston is at the bottom of its bore. This simply means the smaller the crankcase volume, the better it will work as a scavenge pump.

One of the problems with two-stroke engines is that the fuel/air mixture lubricates the connecting-rod and crankshaft bearings. Gasoline is not a good lubricant, so oil is added to the mixture.

This type lubrication would be fine except that oil lowers the octane rating of the fuel. Oil causes problems in a naturally aspirated engine, but they are twice as bad on a turbocharged one, where octane rating is critical.

One way around the octane-rating problem is to use injection mist lubrication in the crankcase, and hope the oil vapor is not carried into the cylinder with the air charge. A more positive answer to octane lowering is to not use crankcase scavenging. On two-stroke engines over 1000 HP, scavenging can be done by driving a turbocharger mechanically until the engine develops enough energy to self-sustain.

This method is not practical on small engines, so a scavenge pump must be provided. You may remember from Chapter 1 that several types of mechanically driven compressors are discussed. All of these have been used at one time or another to scavenge two-stroke engines, Figure 13-5.

In addition, the re-entry compressor, sometimes referred to as a *drag pump,* has also been used for this purpose.

Right away the question arises, "If I have a high-pressure mechanically driven scavenge pump, why can't I use it to supercharge the engine?" Except in the uniflow engine, much of the charge will blow right out the exhaust port and no supercharging will take place.

If a turbocharger is added to the system, back pressure caused by the turbine will retain most of the high-pressure charge in the combustion chamber. When such a setup is used, the turbocharger compressor discharge usually feeds directly into the scavenge-pump inlet, Figure 13-6.

Other Considerations—When calculating engine flow for compressor matching, remember that the two-stroke engine has a power stroke every revolution. Consequently, at the same speed it uses twice as much air as a four-stroke engine of the same displacement.

Another thing to remember is most two-stroke engines have no oil pump. A separate lubrication system—oil pump, reservoir, filter and cooler—must be provided when a turbo is installed.

Traditionally, naturally aspirated two-stroke diesel engines have had better specific output than naturally aspirated four-stroke diesels from both the size and weight viewpoints. The turbocharged four-stroke diesel not only decreased the size and weight advantage, but in some cases reversed it.

Turbocharging Roots-Scavenged Engines—It seems an easy thing to add a turbocharger to a Roots-scavenged engine, Figure 13-6. But unless the engine is specifically designed for both the Roots blower and the turbocharger, the results can be disasterous.

A Roots blower pumps a given amount of air per revolution, but the engine uses a different amount. The amount of air from the Roots blower is always greater than that used by the engine. The difference is accounted for by the air being compressed in the intake manifold before it enters the cylinder.

When a turbocharger compressor is added to the system, it seems logical that the Roots blower will be relieved of some of its duties. Therefore, it should use less crankshaft power. Not so!

Regardless of the pressure delivered by the turbocharger compressor, the Roots blower will still try to run at the same pressure ratio as before. It might even use more power.

In addition, a few more problems arise. The rotors in a Roots blower run with very close clearance in the housing. In a non-turbocharged, Roots-scavenged engine, the rotors continually see fresh cooling air. Therefore, they do not get hot enough to expand and rub on the housing.

If 200F air comes from the turbocharger compressor, some of this cooling is lost. The rotors will probably rub unless rotor-to-housing clearance is increased.

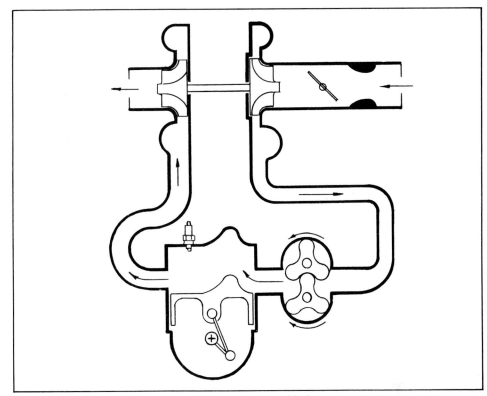

Figure 13-6—Two-stroke engine with turbocharger added to scavenge pump

Also, the oil seals at the ends of the rotor shafts do not see much pressure on the non-turbocharged engine. This of course changes when a turbocharger is added. The seals must then be replaced by a type that can withstand the compressor discharge pressure.

To reduce the power required by the Roots blower, the speed of the rotors may be reduced by changing the drive-gear ratio. With the turbocharger added, the Roots blower is really only needed for starting and light loads. The turbocharger does the major part of the supercharging at high loads.

SUMMARY

Two-stroke engines have been supercharged successfully to produce more than twice their naturally aspirated output. In fact, the potential for a two-stroke engine should be as high as for a four-stroke.

14 High-Altitude Turbocharging

Originally developed for Can-Am racing by Chevrolet, McLaren Cars and Reynolds Metals, Thunder Engines is continuing development of engine for aircraft using twin AiResearch TH08-70A turbochargers and liquid-to-air intercooling. Bendix direct-port fuel injection and twin-magneto ignition is also used. Engine produces 700 HP at 4400 rpm with 48-in.Hg manifold pressure.

Helio Super Courier uses two Rajay turbos on an Avco-Lycoming GO-480 engine.

Turbocharging aircraft engines is a very specialized science, not only because of the unique problems but also from a safety viewpoint. A turbocharger or engine failure in a ground vehicle is annoying, but it's usually not catastrophic. Such a failure in an airplane—particularly a single-engine airplane—will normally cause the airplane to come down.

Aircraft Certification—In the U.S., every effort is made by the Federal Aviation Administration to be sure a turbocharger/engine system is not overstressed. Any controls that are used must be *fail-safe*. Fail-safe in this case means a control failure will not cause the turbocharger to overspeed and destroy itself or the engine.

If you do not have previous test information, you must obtain supplementary certification from the FAA. The cost of building a dynamometer cell with high-altitude capabilities and running the tests for FAA certification will probably cost over $250,000. If this is not a deterrent, go to it.

Turbochargers used on aircraft in-

stallations may *look* identical to those used on farm tractors or automobiles. Quality control, however, is quite different.

Manufacturers must be able to prove that correct materials were used, heat treatment was done under controlled conditions and machining operations checked with regularly calibrated instruments. In addition, special operations such as Zyglo or Magnaflux must be performed on all critical parts.

These are the reasons aircraft turbochargers are so much more expensive than industrial units. This is also why aircraft turbochargers should be repaired or overhauled only by licensed shops using parts from authorized manufacturers.

Because of these special requirements, I don't recommend that owners try to turbocharge their airplane engines. Kits certified by the FAA are available for many popular light planes. If a kit is not available for your airplane, it will be less expensive to sell the plane and buy one for which a kit is available.

Other High-Altitude Applications—After this discouraging introduction, you might wonder why I include a chapter on high-altitude turbocharging. The reason is that many engines must be operated at relatively high altitudes, although they never leave the ground.

Turbocharged engines have done very well in the Pike's Peak Hill Climb and the Bonneville National Speed Trials. Marine engines operating on Lake Tahoe, for instance, suffer a considerable loss of horsepower unless turbocharged.

This chapter covers three areas regaining horsepower on engines operated only at high altitude; maintaining horsepower on engines operated between sea level and high altitude; and overcoming the problems with an engine already turbocharged at sea level when it is operated at high altitude.

REGAINING HORSEPOWER AT HIGH ALTITUDE

A naturally aspirated engine will produce power in direct proportion to

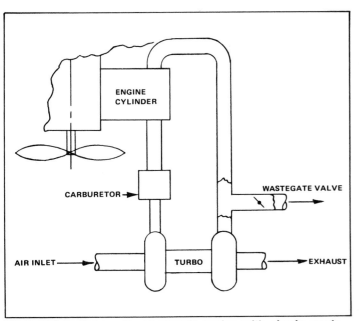

Figure 14-1—Schematic of aircraft engine with simple waste-gated turbocharger

Piper Apache uses single Rajay turbocharger on each Avco-Lycoming O-360 engine to maintain sea-level power at altitude.

the density of the intake air. At sea level, air has a density of 0.0765 lb per cu ft. At 10,000 feet altitude, the density drops to 0.0565 lb per cu ft.

This means an engine that delivers 100 HP at sea level will deliver

$$100 \times \frac{0.0565}{0.0765} = 73.9 \, HP$$

at 10,000 feet.

An airplane engine that uses a turbocharger to regain the power lost at high altitude is referred to as a *normalized engine*. A normalized engine usually has a waste gate to pass all of the exhaust gas at sea level, Figure 14-1. Consequently, no turbocharging takes place at sea level.

As the engine starts to lose power with increased altitude, the waste gate gradually closes, either manually or by an automatic control. The turbocharger then compresses the inlet air to sea-level pressure. This allows the engine to deliver essentially sea-level horsepower.

The engine continues to develop sea-level horsepower up to an altitude where the waste gate is completely closed. At this point, called *critical altitude,* all the exhaust gases pass through the turbine.

When the airplane climbs above critical altitude, the engine will start to lose power. The turbocharger can no longer deliver air at sea-level pressure.

Critical altitude varies with the engine and the turbocharger used. It

is easy to attain a critical altitude of 15,000 feet. With proper intercooling, turbocharged engines have been able to deliver sea-level naturally aspirated power at 40,000 feet.

Altitude requirements dictate the amount of manifold pressure put out by a turbocharger. A vehicle used only at a certain altitude should have a turbocharger matched to deliver 15 psia to maintain sea-level power. If this engine must operate over a broad speed range, then a boost control should be used to maintain 15-psi intake-manifold pressure. Controls are covered in Chapter 10.

If the engine is to be operated at several different altitudes, an aneroid-type control should be used. An aneroid sensor is altitude-correcting, so it will prevent overboosting at lower altitudes.

Supercharging at Sea Level—The above are special cases, where no more than sea-level naturally aspirated power is desired. In most cases, the engine is supercharged at sea level, with the system designed to have as little loss as possible at high altitude.

This type of supercharging can be done in several ways. The specific approach depends on the importance of maintaining full power at altitude.

Figure 14-2 shows how naturally aspirated engine output is reduced as altitude increases. Engine output decreases in direct proportion to air density.

If an engine is turbocharged at sea

level with a free-floating turbocharger, it will also lose power as altitude is increased, Figure 14-3. The power loss will not be as rapid as with a naturally aspirated engine.

With the turbocharged engine at wide-open throttle, *gage pressure* will remain almost constant regardless of altitude. This means *absolute pressure* drops as altitude increases.

As an example, if the engine is turbocharged to 10-psi gage pressure at sea level, it will still be turbocharged 10-psi gage at 10,000 feet.

At 10,000 feet, however, ambient air pressure will have gone from 14.7 psia to 10.15 psia. With 10 psig, the absolute intake pressure was 24.7 psia at sea level. But at 10,000 feet, absolute intake pressure pressure is only 20.15 psia.

The naturally aspirated engine will develop only 74% of sea-level power at 10,000 feet. The free-floating turbocharged engine produces

$$\frac{20.15}{24.7} = 81.5\%$$

of sea-level power, Figure 14-4.

A turbocharged engine with the same sea-level power as a naturally aspirated engine will perform substantially better at altitude. A free-floating turbocharger can compensate for at least part of the power loss at altitude.

Figure 14-4 shows the comparative performance of a naturally aspirated engine and a free-floating turbo-charged engine. If both of these en-

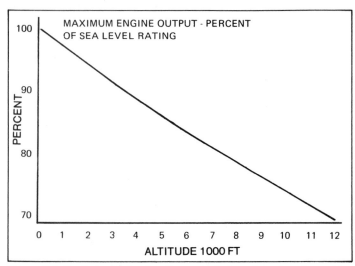

Figure 14-2—Output of naturally aspirated engine at various altitudes

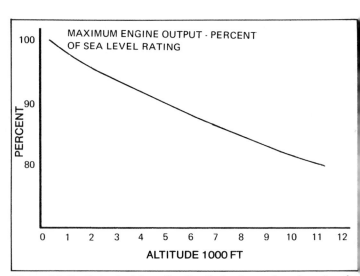

Figure 14-3—Output of free-floating turbocharged engine at various altitudes

Piper Lance aftermarket kit uses two Rajay 315F10-2 turbochargers on Avco-Lycoming IO-540 engine. Separate waste gate for each turbo is operated by a single hydraulic controller.

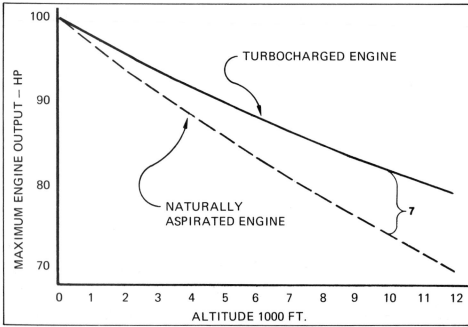

Figure 14-4—Comparison of naturally aspirated and free-floating turbocharged engine output at various altitudes

gines are rated at 100 HP at sea level, at 10,000 feet the turbocharged engine will deliver 9.5% more horsepower than the naturally aspirated engine.

A phenomenon occurs with the turboed engine; gage pressure remains relatively constant with altitude. The turbocharger will gain rpm with altitude. As a rule of thumb, turbocharger speed will increase approximately 2% per 1000 feet of altitude.

There may be applications where there can be no loss of output and the free-floating turbocharger is not acceptable. If the engine is used strictly at a given altitude, the turbine housing may be replaced by one with a smaller A/R. This will increase the rotor speed and boost pressure.

I don't know of an exact method to determine the new A/R to retain sea-level performance. Trial and error should be used until you've established enough "feel" to predict the exact size in advance.

Temperature Effect at Altitude— Depending on the engine and the amount of supercharging at sea level, the engine should be able to produce sea-level power up to at least 16,000 feet. Even if boost pressure is 15 psig at sea level, the lower temperature at high altitudes will help compensate for the "thinner" air.

At sea level the compressor will be producing 2:1 pressure ratio. Intake-manifold pressure will be 30 psia.

At 16,000 feet, ambient pressure is 7.95 psia. To maintain a manifold pressure of 30 psia the compressor must put out 3.78 pressure ratio. This does not seem possible with a compressor such as shown in Figure 14-5.

This map, however, is based on 60F. As air-inlet temperature goes down, a centrifugal compressor produces the same pressure ratio at a lower speed. The exact ratio is:

$$\sqrt{\frac{\text{Standard Temperature Absolute}}{\text{Inlet Temperature Absolute}}}$$

The average ambient temperature at 16,000 feet is about 0F, so this instance is:

$$= \sqrt{\frac{520}{460}}$$

$$= 1.06.$$

Figure 14-5 indicates a turbocharger speed of about 125,000 rpm to produce a 3.78 pressure ratio at 60F. At 0F the actual speed will be approximately

$$\frac{125,000}{1.06} = 117,900 \text{ rpm},$$

which is possible without overspeeding the turbocharger.

CONTROLS & INTERCOOLERS

Other problems at high altitude may prevent the engine from delivering sea-level power. The small turbine housing required to produce the necessary pressure ratio may cause excessive back pressure. The back pressure may cause problems that might reduce the output of the engine.

High intake-manifold temperature may cause detonation and valve burning. Higher combustion temperatures and pressures may cause a general deterioration of the engine. This is a nice way of saying the engine may come apart at the seams.

Some engines must be operated at several altitudes and still be capable of developing sea-level power. This will require a turbine housing small enough for high altitudes and a waste gate to prevent overboosting at low altitudes. The waste gate may be operated either manually or by a servo motor.

It is possible to maintain sea-level output up to the engine's critical altitude, where the waste gate is completely closed. Above this altitude, the engine will act the same as the free-floating turbocharged engine.

Figure 14-6 shows engine output in percent of sea-level rating vs. altitude for an engine where the waste gate is fully closed at 15,000 feet. Assuming this engine is supercharged to 10-psig boost at sea level, the absolute intake-manifold pressure is 25 psia. To main-tain this at 15,000 feet where ambient pressure drops to 8.3 psia, boost must be increased to:

25 - 8.3 = 16.7 psig.

Above this altitude the boost pressure will remain the same because the waste gate is fully closed.

At 20,000 feet the absolute intake-manifold pressure will be 6.75 + 16.70 = 23.45 psia. This will reduce the engine output to:

$$\frac{23.45}{25} = 0.94$$

or 94% of sea-level power. As in the case of the free-floating turbocharger, this engine will have other limits that determine the maximum altitude at which it can operate.

An engine supercharged at sea level must not only maintain sea-level ambient pressure at altitude but the boost as well. As a result, the critical altitude is lower than that of a normalized engine, even though the power at that altitude is higher.

Intercoolers—High-altitude turbochargers can benefit greatly from an intercooler. The higher the altitude the more the benefit.

A turbocharged engine with 10-psi boost at sea-level will have an intake-manifold temperature of about 175F. If this same engine is to produce about the same power at 15,000 feet, boost pressure will have to increase to 16.7 psi.

This boost increase represents a pressure-ratio increase from 1.67 to 3.00. This pressure ratio will give an intake-manifold temperature of 246F unless an intercooler is used.

The standard temperature at 15,000 feet is 5.5F, so an intercooler with only 50% effectiveness will reduce the intake-manifold temperature to 126F.

Manual Controls—Airplanes often limit intake-manifold pressure with a manually controlled waste gate or throttle. Manual controls are used because an airplane engine is essentially operated as a steady-state application. It is not practical to put manual controls on a vehicle that must be driven up and down hills at various power settings and altitudes.

SUMMARY

An engine operated only at a specific high altitude can be turbocharged to correct for air density without the use of a control.

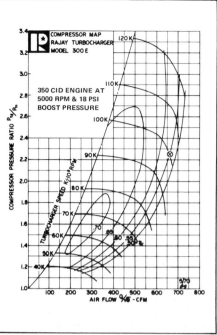

Figure 14-5—Compressor map, Rajay Model 370E

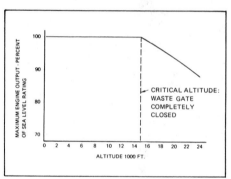

Figure 14-6—Altitude performance of waste-gated turbocharged engine

An engine turbocharged at sea level can be operated between sea level and some high altitude without a control if some loss in horsepower can be tolerated. Percentage loss of power at altitude will be less than if the engine were naturally aspirated.

An engine operated at many different altitudes where maximum horsepower must be maintained must have a control. High-altitude controls are usually some form of aneroid-operated waste gate.

15 Installation Do's, Don'ts & Maybe's

Kinsler Fuel Injection used all the "tricks" when turbocharging Larry Stephens' road-racing Corvette: twin turbochargers, fuel injection, intercooling, tuned exhausts, divided turbine housings and extensive testing.

Installation by P.A.O. Ltd on 3.5-liter Range Rover uses single Rajay turbocharger. Note 90° tee at crossover pipe.

Each chapter in this book covers specific points on turbocharging, but it seems like a good idea to answer the most-often-asked questions in one chapter. This should give you a ready reference.

Many more people are involved in turbocharging now than a few years ago. I have asked several people with considerable experience to pass their knowlege along, so a new enthusiast won't lose interest by making mistakes. Successful people in this field have been kind enough to pass along advice that will save many hours and heartbreaks.

If you have a specific problem, such as with intercooling or lubrication, refer to that chapter and refresh your memory. When I work on a particular item, like a carburetor, I find it extremely helpful to have a book on that carburetor. I'll keep it open to the

page covering the item on which I'm working.

The items covered in this chapter are based on the experience of these experts. In some cases, they may contradict each other—or me. This is natural, because things that are good for one engine are not necessarily good for another. This is particularly true with an air-cooled versus a liquid-cooled engine, or an in-line versus a V-type or opposed engine.

To begin with, some basic suggestions from Doug Roe:

BUYING USED TURBOCHARGERS

"If you are buying a used turbo, follow a few suggestions or that investment may cost you the entire effort.

"Shop around to find a turbo designed for an engine of similar displacement and application. There i ome latitude, but a too-big turbo amps power response. One too small ay be short-lived and restrict mid-nge and top-end performance.

"Keep in mind that automotive-ype units rotate at speeds up to aproximately 150,000 rpm during high utput. Using a turbo that is too small ill cause it to overspeed and fail.

"Before making a purchase, inspect he unit carefully. With the carburetor nd exhaust removed, you can reach he impeller and turbine with your inger or a pencil eraser.

"Using mild effort, try to turn hem from either end. They are joined y a common shaft so both should urn together. If each side turns inde-endently there are internal roblems.

"A used turbo relatively free of arbon will turn with mild effort. The il seal drags a bit, so do not expect it o freewheel. With any luck you can ind one that appears good.

"Do not put any hard object, such s a screwdriver, against either impel-er or turbine to try to turn it. One cratch or nick can cause failure at igh revs. For this reason, never crape off carbon.

"For cleaning, use a carbon-dissolving solvent that does not attack luminum—most impellers are luminum. I use carburetor cleaner. Many high-mileage turbos are heavily encrusted with carbon and will not urn as described above. This means a disassembly must be made to remove he carbon.

"Parts are very costly, so at this point add up the possible costs. A Cor-vair exhaust turbine, for example, costs over $100—the compressor im-eller on the boost side is over $40. A

CarTech kit for Datsun L16, L18 and L20B requires no ducting.

gasket, bearing and seal-overhaul kit is over $40. The latter two items will usually be needed.

"If you want to learn by doing the overhaul yourself, get a manual. Turbos are not complicated, but they are precision made. Use caution when working on them.

INSTALLATION

"Once you have a new or good used unit, it is time to plan the installation. Holding the turbo, try it in various positions and you will soon narrow down the place. Keep in mind the following:

● Hood or body clearance
● Routing the exhaust to the unit

● Routing the exhaust from the unit
● Plumbing oil-pressure and oil-drain lines
● Plumbing the compressor outlet to the carburetor or manifold

Intake Manifolds—"The final comment requires further explanation. The only critical part of the entire plumbing job is how the fuel/air charge is directed from the turbo to the manifold—in-line and V-type engines—or crossover pipe—opposed engines such as Corvair and VW engines. In general there are two things to keep in mind:

● It is best to have the turbo slightly higher than the manifold or crossover tube. If necessary, this rule can be

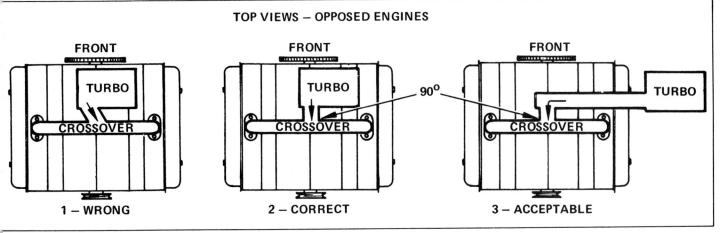

Figure 15-1—Intake-manifold plumbing for turbocharged opposed engines

Brian MacInnes installed this Janspeed kit on his '74 Capri. Kit is designed to fit right- or left-hand-drive cars.

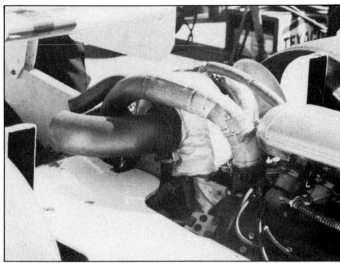

Removed engine cover on Tom Sneva's Indy Car reveals insulating blanket on turbine housing and asbestos-wrapped waste-gate pipe. Blanket on turbine helps maintain exhaust-gas temperature and velocity. Insulated turbine and pipe reduces engine-compartment temperature. Photo by Tom Monroe.

broken without big losses of power.

• Regardless of where the turbo is mounted, never plumb to the manifold or crossover pipe so the charge enters at an angle.

"When aimed wrong, Figure 15-1, heavy fuel particles will tend to continue along the easiest path. In this case the right-side cylinders will run richer. Air is lighter, so it can change direction faster. It will make the sharp turn to the left and cause those cylinders to run lean.

"In short, straighten the path of the fuel/air and let the cylinders have an equal chance at it.

"The same applies when plumbing to a manifold setup on an in-line or V-type engine. You must provide a means of shooting the fuel/air straight in so it is directed evenly to all cylinders.

"A study of any turbo installation will reveal an exhaust pipe from all cylinders plumbed to the turbo. Plan your exhaust to be as short as possible without causing tight bends. Use pipe large enough to ensure free flow."

Exhaust Wrapping—Doug suggests wrapping the exhaust pipe with insulating material to keep from losing any heat in the exhaust system. This also lowers the temperature of the engine compartment.

Exhaust wrapping is standard practice for racing, where fractions of a second mean the difference between winning or losing. On street applications, the installation may cause more trouble than it cures.

George Spears suggests: "In regard to the subject of wrapping the exhaust-to-turbine pipe, we experimented on the V6 Capri and the V6 Mustang and found very little or no gain.

"What we found was a tremendous deterioration of the wrapped pipe. It rapidly oxidizes. In some cases, we found the bends distort and the crossover pipe warps. Once removed, the exhaust system cannot be reinstalled.

"For most automotive applications, wrapping is unnecessary, costly, seriously affects exhaust-pipe life, and generally is not attractive."

In both cases, George was referring to a V6 engine. On any V-type or opposed engine the crossover pipe takes the most punishment.

The engine itself is water-cooled and will not get much over 220F. Therefore, the distance between the exhaust ports on the two banks will not vary much. But the crossover pipe connecting the ports may reach a temperature as high as 1500F.

At this temperature, it will grow about 1/3 inch more than the block if the pipe is made from 300-series stainless steel. With this much growth something has to give. Because the engine block is usually far stronger than the crossover pipe, the pipe will give.

When Rajay first produced their Volkswagen Turbocharger Kit, the pipe from the exhaust ports to the turbine inlet was wrapped with an efficient heat blanket. A few months later, one of these pipes was returned as being defective. Several pieces ha flaked out of the corners nearest th exhaust ports.

This was thought to be a bad piec of tubing until several more heade were returned with the same problem The kit was redesigned and the heade was chrome-plated rather tha insulated. There was no appreciabl performance difference between th two types and the chrome-plate headers had a much better appearanc and life.

Don Hubbard says, "Some peopl say the turbo is operated by the e× haust *heat* of the engine. They ar tempted to insulate the exhaust tube between the engine and the turbo t get more heat to the turbo. This usua ly causes overheating of the exhau tubing and premature failure.

"Exhaust gases actually have velocity and pressure along with th heat that drives the turbo. Usually the insulation used around the e× haust system is to keep radiant he off other engine parts.

"In some Indy cars, insulation ha been used to keep the high-speed a stream from cooling the tubes exce sively because the state of tune of th engine requires every little bit fro the turbo for response and power an the balance is critical.

Intake Connections—"Connectio between the compressor outlet an the intake manifold should not b overlooked. A leak at this locatio will not only cause a loss of power an rough idling but can be a fire hazar as well. This is particularly true on e

Doug Roe turbocharged this 3.0-liter Buick V6-powered Skyhawk for competing at the Bonneville Salt Flats. Engine uses 9.25-to-1 Diamond cylinder heads, Isky cam, reworked Hilborn fuel injectors, twin AiResearch TO4's and air-to-water intercooler. Water from ice chest is circulated through intercooler by crank-driven pump.

Ron Fournier, master metal fabricator and author of HPBooks' *Metal Fabricator's Handbook,* installing shield to protect brake master cylinder from radiated exhaust heat in GTX Mustang. Heat shields protect components from extreme heat in tight spots.

ARKAY Triumph TR-7 kit uses Rajay turbocharger and waste gate, side-draft-Weber carburetor, water injection and VDO boost gage.

Array of instruments in the above Skyhawk supply turbocharger and engine data for on-track development work. Watching gages and keeping car between the lines at 200 mph can keep one busy!

gines with draw-through turbocharger installations.''

Doug Roe suggests, ''When connecting the turbo and your newly fabricated intake stack or crossover pipe, use a good grade, fuel-proof hose and clamps.

''If a modest bend is required, use wire-lined flex hose with a relatively smooth interior. Keep the pipe size in the intake system similar to the compressor-outlet size to get best connection with least irregularity.''

For the connection at the compressor discharge, George Spears recommends red or green silicone hose manufactured by Flexaust or Flex-u-sil. Conventional radiator hose is quickly attacked by gasoline. Paul Uitti has had good results with Arrowhead hoses.

LUBRICATION

When firing up a newly rebuilt engine with a turbocharger, you needn't take any more lubrication precautions than on a naturally aspirated engine. The turbocharger barely rotates at engine idle, so oil used to lubricate the turbocharger bearing during assembly will be adequate until the engine oil passages fill.

There is always the possibility of hooking up oil lines incorrectly. To check, disconnect the oil-drain line from the turbocharger and observe oil flowing from the turbocharger drain after the engine starts; have a clean container underneath to catch it. Oil should flow within 30 seconds if the correct oil viscosity and line sizes are used.

Doug Roe has some advice about the Corvair that might help prevent a turbocharger failure. ''Before starting the engine, connect the oil-pressure line from an oil-pressure outlet on the engine block to the turbo. This should be 1/4-inch steel tubing or a good flexible pressure line with equivalent oil-flow capacity. The line should be capable of handling hot oil.

''As a precaution to possible turbo-bearing damage, pump oil through the turbo before firing the engine. Turn the oil pump with an electric drill. On the Corvair I used a distributor with its drive gear removed.

"Determine how your engine's oil system functions and devise a way to work the pump. Turning the oil pump without running the engine also guarantees immediate oil pressure when you fire the engine. This is a very good practice.

"Run the oil pump long enough to observe a good flow of oil out the drain hose. This will be a small steady stream with low pressure and room-temperature oil. When the oil gets hot and has full engine oil pressure against it, the drain will flow more oil. Doublecheck everything and run the engine for leak checks."

Oil Drains—Don Hubbard, as well as the others, knows the importance of the turbocharger oil drain. He suggests the following: "The turbo needs a good direct drain to the engine. Make the drain tube to the engine approximately 3/4- to 1.0-inch inside diameter.

"If the drain goes to the oil sump, the sump drain should never be covered with oil. It should terminate in a position *above* the oil level and be protected from oil thrown by the crankshaft and rods. The crankcase should be well vented to prevent high crankcase pressures due to blowby.

"Most oil-leakage problems in the turbo can be traced to poor oil drains. Drains must *always* slope down and have no low spot to trap oil in the drain tube."

This item is discussed in the lubrication chapter. No matter how often it is talked about, some people still seem to think oil can drain uphill.

INTERNAL ENGINE MODIFICATIONS

People who have run turbocharged engines frequently say the valves float at a lower rpm than when the engine was naturally aspirated. Some attribute this to the supercharge pressure on the manifold side of the valves, causing them to stay open. Here's their reasoning:

Consider an engine with 2.25-inch intake valves that is boosted to 10 psi. Intake-manifold pressure exerts approximately 10 psi x 3 sq in. = 30 pounds against the backside of the *closed* valves. (The effective area of the backside of a 2.25-inch valve is

TRW supplies dished pistons for reducing compression ratio. Small-block Chevy *Turbo* piston L2441F drops compression to 7.65 to 1 with 76cc heads. Turbo pistons are also available for the big-block Chevy (L2453F) and 3.8L Buick V6 (L2481F).

Bronze-silicon valve-guide inserts have excellent wear properties, even in turbocharged engines.

about 3 sq in., considering the area masked by the valve seat and the valve stem.) The fallacy is, the force against each intake valve doesn't exist until it is closed. Additionally, there's the rising pressure in the combustion chamber as the piston travels up the bore. The point is, although the pressure is higher on the manifold-side of the valves, it's also higher on the combustion-chamber side. The differential pressure is essentially the same.

Valve Springs—Doug Roe has a few comments: "If stock valve springs at stock installation heights are used in a turbocharged engine, valve float may occur at lower rpm than was the case with the naturally aspirated engine. This can probably be attributed to the following causes:

• There is a tendency to use the full rev-limit capabilities of a turbocharged engine. The exhilarating experience tends to generate driver enthusiasm, which may not be accompanied by tach watching to ensure the critical rpm is not exceeded.

• Valve springs in used engines may already be *tired* and not giving specified seating pressure. Installing a turbo on such an engine—and then attempting to use full performance and rpm potential will often bring about valve float.

"Continued use of high rpm weakens the valve springs in a hurry. Stronger springs or a lower installed height—either of which increases

valve-spring pressure—will raise the rpm limit.

"If the springs are shimmed, it is important to observe the usual precautions to prevent coil bind with the valve fully opened.

"An engine does not have to be very old to have the 'tired-springs' syndrome. If such an engine is turbocharged, consider changing the valve springs when the turbo is installed. Even a new engine can have tired springs, particularly if it sat with the heads assembled.

"If excessive valve-guide wear is encountered on a turbocharged engine—and this is not usually a problem—the K-line method of bronze bushing the guide bores can be helpful.

"Bronzewall Guide rebuilding by the Winona Method is another repair. It uses aluminum-bronze wire wound into the guides. Bronze has more lubricity than guides made from the Detroit Wonder Metal—cast iron.

"Also, the valve-stem seals can be left off the exhaust valves/guides. The exhaust valves will get slightly more lubrication to offset any effects of higher exhaust-gas temperature and increased back pressure.

Head Gaskets—"Head-gasket problems may occur on some engines, especially as boosts get into the 13- to 15- psi region. If problems occur, they can be cured by *O-ringing* the block.

"O-ringing requires cutting grooves in the block-deck around the outside of each bore. Soft copper wire

Compressor-inlet provides solid clue as to flow capacity. Left is original Corvair B-flow; right is Corvair E-flow kit by Crown Manufacturing.

is installed in the groove with a portion of the wire extending above the surface. This increases gasket pressure around the combustion chamber and greatly reduces any tendency to blow gaskets."

Head gaskets incorporating O-rings are also available for Buick V6 and Chevrolet V6 and V8 engines. No machining of the head or block is required to use these gaskets.

Datsun uses gas-filled O-rings to seal cylinder pressures on their Turbo 280ZX engines. The rings drop into steps at the top of the cylinder bores and stand higher than the gasket surface of the block. When the head is torqued in place, the gaskets are compressed against the head-sealing surface.

"Head-gasket problems are rare on air-cooled engines that *spigot* the cylinders into the heads. In these cases, gaskets are retained around their edges by the head design. These have little tendency toward failure, even at high boost pressures.

"VW and Porsche cylinders seal directly against the aluminum heads to provide a very effective seal. This assumes the cylinder lengths, spigot depths in the heads, and cylinder bases on the crankcase halves are correctly machined.

"Correct head-bolt/nut torque is essential. The usual practice of running the engine until it is well warmed up and retorquing after cooling is highly recommended.

Exhaust Valves—"Better-quality exhaust valves are used by some builders when turbocharging. Turbocharged Corvairs, for instance, had Nimonic exhaust valves as standard parts. As valve diameter is increased and/or boost pressure is raised, the need for better-quality exhaust valves increases.

UNDER THE HOOD

Clearance—"It is very important to make sure that the turbo does not hit anything when the engine moves on its mounts or the chassis flexes. Pay particular attention during acceleration or braking or on rough roads, as in off-road racing. Several turbo failures on a single car were finally traced to the fact that the turbo would hit a rollbar brace under certain conditions. It appeared to have plenty of clearance with the engine and car at rest.

"Remember that turbos rarely need expensive carburetors, intake manifolds or exhaust headers. Eliminating the cost of these items alone can often offset the cost—even for a pair of turbochargers. Obviously, if you are intent on *buzzing* the engine to higher rpm, it should be treated to a detailed blueprinting job."

George Spears provided the following tips on fabricating exhaust pieces: "We experimented with various thicknesses of plate for the flame-cut flanges, and found that 3/8-inch is a happy compromise. 1/4-inch has a tendency to warp during fabrication

and welding, whereas 3/8 inch will not. We discovered that 0.083-inch-wall (14-gage) tubing is adequate. Anything thicker is even better."

Underhood Heat—George offered some good suggestions on protecting engine components from hot exhaust pipes. "I think a question most frequently asked concerns how close the hot exhaust piping can be placed next to heat-sensitive objects.

"Our Mustang II Kit has its turbine discharge approximately 3/4 inch away from a 14-gage (0.0747-in.-thick) metal heat shield. Electrical wiring is immediately behind this heat shield and, is actually touching it in some cases.

"We have not experienced any problem with the wiring. We have placed thermocouples against the front side of the heat shield and generally can only measure about 200F. Based on this experience, I think we can say that a 14-gage-steel shield placed 1 inch away from any of the hot metal parts, with at least a 1-inch air space behind it would give adequate protection from the heat. Shields should be placed near any critical parts such as electrical wiring, master cylinders, etc.

"This might be useful information to an individual who has absolutely no idea of the heat problems."

HIGH-BOOST INSTALLATIONS

Jim Kinsler has done a great deal of work with fuel injection and intercooling on turbocharged racing engines. Regardless, many of his recommendations apply to street engines.

Intercoolers—Jim Kinsler is a strong advocate of intercoolers, especially on gasoline engines with more than about 7-psi boost. He commented, "An intercooler will lower the temperature of the charge going into the engine enough that:

• The engine will gain 15 to 20% more power.

• The engine will be less prone to detonation.

• The lower inlet temperatures will also lower the exhaust temperature about the same amount. Lower exhaust temperatures will give longer exhaust-valve and turbo life.

"For engines running high boost pressure, especially when using gasoline, it is critical to obtain the

COMPUTER-CONTROLLED ENGINES

If you have one of the new computer-controlled engines, such as GM's Computer Command Control (CCC) or Ford's Electronic Engine Control (EEC), your car won't take kindly to turbocharging. This is because the computer is programmed to make ignition and carburetion changes based on intake-manifold pressure, exhaust oxygen content, and a myriad of other inputs. Engine calibration is so exact that changing something as simple as the muffler can foul up engine performance. And, a turbocharger will really foul up things. Ignition-timing and fuel requirements are significantly changed.

Doug Roe discovered this problem the hard way when he attempted to turbocharge a 1982 Cadillac Cimmaron. Because he had no way to reprogram the computer, the car ran very badly after he installed the turbocharger. He realized it would be too time consuming to convert the engine-control system so it would be compatible with the turbocharged engine. Doug also knew that aftermarket entrepreneurs were hard at work developing a device that could be used to test and recalibrate the Engine Control Module (ECM)—the "brain" used in the

GM's CCC system since 1981. So, he removed the turbocharger, packed it in a box and waited for the tuning device.

Digital Automotive Systems, Inc., 7201 Garden Grove Boulevard, Garden Grove, CA 92641, offers a small digital-electronic tester for GM cars. (Testers for other brands are in the works.) The test unit plugs directly into the ECM. Although it's primarily a service tool that can monitor all engine-control systems, troubleshoot senders and solenoids, and read out trouble codes, the tester also has the ability to change, among other things, the spark curve and fuel metering. This can be done with the car on a chassis dynamometer or on the road!

Therefore, all that's required to turbocharge a computer-controlled GM engine successfully is to first install a manifold absolute-pressure (MAP) sensor from any turbocharged GM engine—it's range is 60 in.Hg versus 45 in.Hg for normally aspirated engines. To control detonation, you could also install a detonation sensor. Once these sensors and the turbocharger are installed, use the tester to reprogram the ECM to make it compatible with the turbocharger or any other engine modification.

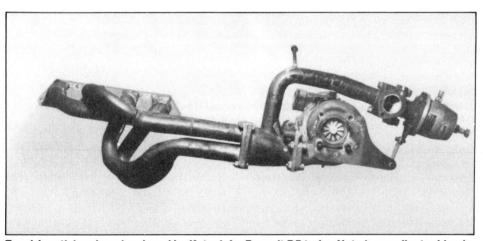

Equal-length headers developed by Katech for Renault R5 turbo. Note how collector blends smoothly from primary pipes to turbine inlet. Fritz Kail of Katech stresses that intake and exhaust tuning is just as important on a turbocharged engine as on a naturally aspirated one.

lowest possible temperature for the charge air entering the cylinders.

"It is much more critical to get a good cooling air supply to the intercooler than to the radiator. Also, the ducts from the turbos to the intercooler and from the intercooler to the engine should be kept short to reduce system volume.

"In a front-engine car, the intercooler usually takes the place of the radiator. Lengths of the radiator hoses are not critical, so radiators are often

placed in the rear fenders of the car. Air enters the radiators through ducts on top of the fenders. Low pressure behind the car helps draw air out.

Intake & Exhaust Tuning—"One of the biggest misconceptions is that ram tuning is not important on a turbocharged engine. Through extensive testing, I've found that a turbocharged engine responds to a tuned inlet and exhaust system and large-port cylinder heads as much as a naturally aspirated engine. The engine also needs much

more cam than most people think. Don't strangle the engine!

"One of the best examples of this was a 2.0-liter BMW that Gary Knudsen at McLaren Engines developed for Can-Am racing. The engine produced 540 HP on gasoline with Mack air-to-air intercoolers and a simple log-type exhaust manifold.

"Gary replaced the log manifold with tuned headers and picked up 60 HP. When he adjusted the cam specs to take advantage of the better exhaust, the engine produced 640 HP on the dyno and 600 on track! When Gary installed the tuned headers, throttle response also improved tremendously.

"Gary went on to experiment with axially divided turbine housings, an available option from most turbocharger manufacturers. He found that if the exhaust pipes were connected to the separate sides of the divided exhaust housing so the exhaust pulses entering each side were evenly spaced, throttle response improved even more. Though this arrangement didn't add power, 'turbo lag' was all but gone."

Most V8 engines don't fire evenly on the right and left banks. That is, the firing order doesn't alternate from right to left. On a dual-turbocharger installation, if you want to have even exhaust-pulse spacing, you can't simply run one turbocharger off the right bank and the other off the left.

The only way to get evenly spaced pulses into the two sides of each turbine housing is to bring pipes from the right and left sides of the engine to the housing. The pipes are usually routed behind the engine above the bellhousing. This keeps the pipe shorter than they would otherwise be.

Kinsler has seen these same results over and over in the systems he has built. Some of the world's quickest road-racing cars have used this arrangement to improve response out of low-speed turns.

When fine-tuning a new system, Kinsler uses three exhaust temperature probes: one at the exhaust port, one at the turbine-housing inlet and one at the turbine outlet. A temperature rise between the exhaust port and the turbine inlet indicates fuel burning in the exhaust pipes. This generally indicates not enough spark advance, though it could also show too much valve overlap or excess fuel.

Kinsler stresses that, "Spark advance is absolutely critical in a turbo engine. Too much advance and the engine detonates. Too little advance and you burn up the exhaust valves, pipes and turbocharger."

Temperature drop from the turbine inlet to the outlet, combined with the pressure drop across the housing, is used to determine the overall efficiency of the turbocharger system.

For exhaust-system material, Kinsler prefers type-321 stainless, or at least 316 stainless. On a high-output turbo engine, either will last, even at maximum operating temperatures; or "forever" on a street engine.

Fuel Injection—Fuel should always be added after the turbo—that's why Kinsler recommends fuel injection.

"On an intercooled installation, adding fuel before the intercooler will cause fuel puddling and fuel lag, the major cause of 'turbo lag.'

"It is important to add the fuel just as the air enters the combustion chamber for several reasons:

● If you put fuel into the air before it enters the turbo, system efficiency decreases. The temperature of the charge air as it enters the engine increases, causing poorer performance.

● So-called 'turbo lag' is often really *fuel* lag. Injecting the fuel at the engine intake ports not only sharply increases the throttle response, but also eliminates any problems with fuel distribution."

Kinsler makes and sells constant-flow fuel-injection systems, and notes that these work well on naturally aspirated or mechanically supercharged engines. However, he has found that the constant-flow system cannot meet the tougher requirements of a turbo engine, especially for road racing, even when the systems are equipped with a boost sensor.

"The problem with the constant-flow system is that there is no way to compensate for the boost pressure at the nozzles *through the whole operating range.* A particular setting may be perfect at one boost pressure, but will be too rich or lean at others.

"A positive-displacement, timed-injection system works well when equipped with a boost sensor. Lucas and Bosch mechanical timed-injection systems are two adaptable systems that work well. Boost pressure working against the nozzles does not affect the metering of these systems at all."

TURBOCHARGER KITS

Bob Helvin in England has been associated with turbocharging for many years. Recently Helvin has worked closely with most of the kit manufacturers in Europe as well as England.

Many of their programs are similar to those in the United States, but some are peculiar situations—such as no speed limit in Germany. Because of this, I asked Bob to offer a few hints, which turned out to be good advice no matter what language you may speak.

"In Europe, hardly a week goes by without a turbocharged car being announced. Some are by major manufacturers and some are by the fast-growing fraternity of turbocharger kit-manufacturers.

"One has only to visit the major international motor shows in Europe to realize the word *turbo* is being applied to anything. The word has a magic ring to it, unlike any other product I have known during my twenty years in the motor industry.

"In the 1980s, the turbocharging industry is growing rapidly. In the aftermarket, many companies that previously tuned by conventional methods—different carburetors, cylinder-head modifications, and so forth—have adopted turbochargers. These companies are learning rapidly how to overcome the traditional disadvantages of turbocharged cars—poor low-speed acceleration and turbo lag.

"These companies are now producing some exciting, reliable and durable installations. In each country, prices compare favorably with conventional tuning.

"There are two schools of thought in Europe. One is that the engine should be modified—even blueprinted—then given a high boost of 9 psi or more. Modifications include forged pistons, strong but light connecting rods, strengthened crankshaft, larger-capacity oil pump, modified cylinder heads and additional engine cooling.

"This particular school is concerned mainly with getting as much horsepower as possible from the engine, sometimes in the region of 125 HP per litre or more, and is concerned mainly with producing low volume specialist kits.

"The other school is more conservative and by far the larger. These people are working at producing turbocharger kits for normal road use.

This school considers a 30 to 40% horsepower increase to be sufficient when it won't compromise engine life.

"This school also believes that bolt-on turbocharger kits should be just that—bolt-on. A competent mechanic should be able to install the kit within a few hours.

"Often the only engine work involved is to lower the compression ratio. This is usually achieved by machining the combustion chambers, lowering the piston crown, or installing a *decompression plate(s)*—spacer(s) between the head(s) and block.

"I believe that lowering the piston crown is very suspect. Knowing the piston crowns will be subjected to higher temperatures, I prefer that they not be machined. If you are tempted to modify the piston crown, I suggest installing new low-compression pistons. This modification is only necessary on certain engines.

"Everything else is bolt-on. A maximum of 5- to 7-psi boost is used by these companies, which gives excellent performance while retaining driveability.

"The Rover SD1 is a perfect example. With moderate boost, the car goes 0-60 mph in 6.6 seconds. Turbocharging gives the car a top speed above 140 MPH and sufficient torque to allow the driver to pull away smoothly from 1000 rpm in top gear.

INSTALLATION GUIDELINES

"So much for the turbo. What about the rest of the car? To start with, the engine must be in good condition. I've found that a good guide is to check that the compression pressures are equal and within about 5% of manufacturer's specification. Also check that the oil pressure is within 10% of specification.

"In general, any engine with more than 20,000 miles on it should not be turbocharged. Also, the transmission, chassis, brakes and steering should be checked to be sure they will stand the extra performance.

"How critical these points are will of course depend upon the specifications of the standard vehicle. Most of the kit manufacturers keep this in mind. Many will offer other modifications tailored to a specific vehicle, either as part of the kit or as an accessory.

Marketed by American Turbocar Corp., Camaro/Firebird turbo kit by Gale Banks is truly complete. Kit includes dual-turbocharged 500- or 600-HP small-block Chevy, Nash 5-speed transmission, 9-inch-Ford rear axle and heavy-duty driveline and suspension components.

Turbo Tom's toy is this L20B-powered road-racing 510 Datsun. Turbocharger draws through a four-barrel Holley topped with a K & N air filter. Photo by Doug Roe.

"For example, turbocharging some cars may enable them to exceed the speed rating of the tires. Or it may be that the car needs larger-diameter wheels or wider rims. Some kit suppliers offer this equipment as part of the package, or as an option.

"It is obvious that the extra power will be used. The higher speeds place extra demand on the braking system.

"To eliminate brake fade, brake pads and shoes of a much harder material can be used. If solid rotors are used or the rotors are too small in diameter, ventilated discs can be installed.

"One could go on, but I merely use these examples to illustrate. Turbocharging does not always end by installing the turbo and driving the car away.

"Listed below are some of the things I have come to consider as requirements over the last couple of years:

Simplicity—"Keep the installation simple—somebody once said simple is beautiful—from a cost, performance and installation standpoint. European cars sometimes make it difficult to install the turbo in the ideal position because of limited underhood space. Some of the plumbing becomes a nightmare, not only from an installation point of view but for accessibility.

Reliability—"The kits must be reliable—one kit manufacturer told me, *idiot proof.* Much higher speeds are attained from smaller engines in Europe. In Germany, for example, there is no speed limit. Vehicles are driven fast and turbocharged vehicles are driven fast for long periods.

"A carefully matched turbo with a waste gate or similar device to limit boost puts no additional stress on the engine. That's assuming the manufacturer's recommended engine speed is not exceeded.

Lubrication—"With the additional heat generated by smaller engines, it is often found necessary to use an oil cooler. In many instances the oil cooler is supplied as part of the kit.

Intercooling—"Intercooling is becoming more popular. For example, in Holland, Hi-Tuned is using intercooling extensively. But Hi-Tuned uses boost pressures of 9 psi and upward in some cases.

Underhood Heat—"To help keep the turbocharger cool, the ideal mounting is in the air stream. This is not always possible, particularly on draw-through applications.

"The actual positioning of the turbocharger can lead to problems in other areas if some precautions are not taken. Brake lines and master cylinders should be shielded.

"I know one instance where a clutch cable was located too near the turbine housing. The car ran fine until it became increasingly difficult to shift. The clutch-cable conduit finally melted. The owner of the car now changes cables every few thousand miles as a matter of course. Even with all their technology, Saab had to recall some cars because of melted electrical cables.

Fuel Requirements—"Ensure that good-quality fuel is used. I attended a journalists' day at a race track last year. I took along a Datsun 280ZX and a Rover SD1, both turbocharged by Janspeed.

"The cars ran very well until they needed fuel. When we used the fuel from pumps at the track, almost immediately both vehicles began to detonate. This continued until we left the track and bought some decent fuel.

Air Filter—"Air-filter size is quite important. I've seen several good installations ruined by very small but attractive air filters. The engine ends up strangled. I've even seen some installations where a piece of wire net was placed over the end of the compressor inlet. The net was drawn into the compressor.

Boost Gages—"There is a big argument for and against boost gages. I believe in keeping distractions to a minimum. In a car used for day-to-day driving, the fewer instruments there are the better.

"People tend to get hung up on boost gages. If a gage shows 6.5 psi instead of 7 psi, some people will be dissatisfied with the whole installation. The claimed readings are not indicated, even though the gage may be the problem.

"I would prefer to see a warning device—light or buzzer—that operates whenever there is a problem—if the waste gate doesn't open, for example. My whole philosophy is—if there's no problem, leave it alone.

Hardware—"Kit manufacturers are moving away from fabricating manifolds and pipework for volume production. Many now produce high-quality castings. These tend to make the kit look like part of the original engine and not just an add-on.

Spares Availability—"If something goes wrong with either the turbo or an installation, can you get spares? I have found it is important to have a nationwide backup for any kits. Because the turbo is a mechanical component, something can and will go wrong eventually, and spares will be needed quickly.

Maintenance—"Regular maintenance of the oil and air filters and the air-filter ducting are important. Most turbocharger failures can be attributed to the lack of maintenance in these areas. I am not saying the turbocharged engine demands more regular servicing than its naturally aspirated counterpart. It just needs servicing at the recommended times.

INSTALLING THE TURBO

"Before installing the turbocharger, a number of elementary checks should be made.

"Be sure there is no foreign material in either the air-inlet ducting or the intake and exhaust manifolds.

"Check all connections for

Turbocharger on Indy Car is mounted to transaxle rather than engine. This is OK because there is no relative movement between engine and transaxle. Note low-mounted waste gate and divided turbine housing. Photo by Bill Fisher.

tightness. The clamp bolts around the compressor and turbine housings should be torqued to specification, otherwise the turbo may fail and the warranty could be denied.

"Before attaching the oil drain to the turbocharger, crank the engine but don't allow it to fire. Once oil flows from the turbocharger bearing housing, the pipe can be connected and the engine started.

"For the first five minutes or so the engine should be idled and not revved to high speed. Once the engine is shut off, the oil pressure drops to zero, but the turbocharger is still freewheeling. Consequently, turbo and engine life are both prolonged if the engine is brought to idle before turning off the engine. Finally, check all the systems for oil, air or exhaust leaks."

Warning: Do not touch the turbine housing when the engine has been run. It will be very hot and you can be severely burned.

TROUBLESHOOTING

"When a problem develops on a turbocharged engine, the turbo is immediately blamed. Only after replacing the turbo without solving the problem does anyone do any further checking.

"Once removed, the turbo is either needlessly exchanged, repaired or returned to the manufacturer under warranty. Of course, warranty is denied.

"There are a few simple visual checks for the condition of the turbo. These should be done before remov-

Callaway-turboed 320i BMW shows connection of waste-gate discharge to exhaust. Photo by C. E. Green.

ing it from the engine. The following, I believe, cover all the main requirements:

● On the air intake, check for loose pipe connections between the air filter and the compressor. Any gaps here could allow dust into the system.

● Check the duct from the turbocharger to the engine intake system. A loose connection could cause low power and a high-pitched noise.

● Remove the intake and exhaust pipes from the turbo and check the compressor and turbine wheels for impact damage from foreign objects.

● Check for evidence of the wheel contacting the housing walls. This would indicate bearing failure. Rotate the wheel by hand—it should rotate freely without touching either housing. Note: There will always be some side-to-side movement but this movement is built into the turbo.

● Check the exhaust manifold for cracks or loose nuts and bolts. Also check that the turbo fastening bolts are tight. Exhaust leakage here will cause noise and loss of power.

● Check for oil loss through the turbocharger seals. If oil is evident in the exhaust system or exhaust gas, check that the oil-drain hose is not bent or corroded. Any oil-drain restriction can cause oil loss through the turbocharger seal.

"If these simple visual checks fail to reveal any obvious faults, the problem is almost certainly not with the turbo."

16 Tractor Pulling

Dicky Sullivan of Naylor, Missouri, behind wheel of his 414-CID International. Hypermax-designed three-stage turbocharger system produces 250-psig manifold pressure! Two turbos in parallel are used in the first stage and one each for the following two stages. Photo courtesy of National Tractor Puller's Association.

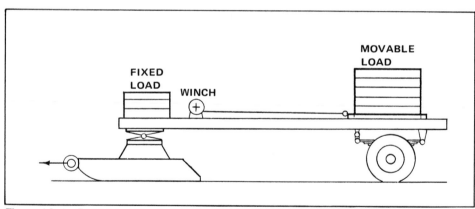

Figure 16-1—Dynamometer sled used in tractor-pulling contests

Tractor Pulling has become a big spectator sport in the United States and Canada. England had joined the sport and I wouldn't be surprised if pulling becomes one of *the* international pastimes.

The National Tractor Puller's Association is the largest of the many sanctioning groups. Annual attendance, as well as purses, are well into the millions at NTPA events.

Because some readers are probably saying, "What is a tractor-pulling chapter doing in a turbocharger book?" let me explain.

Information supplied by the NTPA states that the first recorded tractor pulls were in 1929 in Bowling Green, Missouri and Monsville, Ohio. The tractor started out pulling a sled loaded with stones. As the sled was pulled along the course, a man climbed on it every 10 feet. If it was pulled 100 feet, the sled had the weight of the sled and stones plus the additional weight of the 10 men. This continued until the tractor stalled.

In the early days, that was it. As the years went on, tractors became heavier and more powerful. It was necessary to divide them into classes.

At most tractor-pulling contests there are four divisions: super stock, modified, mini-modified and four-wheel-drive truck. Within each division there are many weight classes.

A modified tractor has any combination of engines, transmissions and final drive. It must have rubber tires and two-wheel drive. Super-stock tractors are farm tractors with a standard block and crankcase of the make and model originally installed. These two classes are further divided into weight divisions of 5000, 7000, 9000, 12,000 and 15,000 pounds gross.

Mini modifieds are V8-powered garden-tractor-size machines weighing 1500 or 1700 pounds. Four-wheel-drive trucks compete in 5500- and 6500-pound classes. These are the fastest growing class in the sport. They may look like the average off-road pickup, but to be competitive they must have more than 1000 HP.

Many other rules cover safety requirements, like SEMA-approved scatter shields and blankets. Rules may be obtained from the National Tractor Pullers Association, 104 East Wyandot Street, Upper Sandusky, OH 43351.

The old method of pulling a sled with men climbing aboard is no longer used for safety and other reasons. The sled has been replaced by a trailer with a skid on the front and wheels on the back, Figure 16-1.

The trailer starts out with a fixed load on the skid end and a movable load on the wheel end. Total weight depends on track conditions and the tractor weight division.

The movable weight starts out above the rear wheels. As the trailer is pulled along the course, the weight is gradually moved forward. Drag is increased by transferring the major part of the load from the wheels to the skid.

The load eventually becomes so great that the tractor bogs down. Distance from the starting line to the trailer hitch is measured to determine which tractor pulled the load the farthest.

Engine size is unlimited in any class, so it's uncommon for a tractor to stall or "power-out." The usual way a run ends is with the tractor at a complete stop with the wheels still turning.

TURBOCHARGING TRACTORS

This brings us to the reason for a chapter on tractor pulling in a book on turbochargers. Turbochargers were first introduced to tractor pulling in the early '60s—long before they were common at Indianapolis or on the Can-Am racing circuit. Some of the first applications were crude.

It wasn't long before the pullers learned which turbocharger worked best on which engine and how much manifold pressure could be used before the engine blew. Originally, 10- to 15-psi boost pressure felt pretty good. It gave the puller a big advantage over the naturally aspirated engines.

Once turbochargers were common, it was necessary to up the boost pressure to win. The limit from a single-stage turbocharger was about 40 psig, but the engines still didn't give up. This is about the maximum pressure run on Indy-type race cars.

Staging—The next step was to mount

Bill Williams of Walsh, Colorado uses three series staged turbocharger on his John Deere 4010. Water is injected between each stage for intercooling. Photo courtesy of National Tractor Puller's Association.

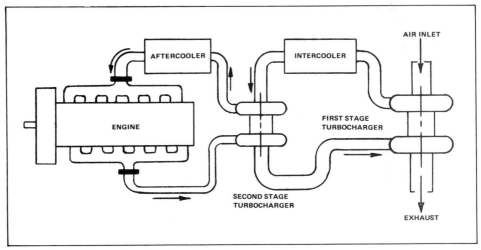

Figure 16-2—Schematic of tandem-turbocharger system

the turbochargers in tandem, Figure 16-2. With this method it's possible to get ridiculous manifold pressures without overstressing the turbos.

With staged turbos, intake-manifold temperature can go out of sight, so intercoolers were added between the two compressor stages. Each run only takes a couple of minutes, so ice water is used as the cooling medium.

In Dr. Alfred J. Büchi's *Monograph on Turbocharging* published in 1953, he discusses the power increase of a diesel engine as a function of charging-pressure ratio.

Dr. Büchi determined that the output of a correctly scavenged and intercooled turbocharged engine continued to increase at pressure ratios as high as 4:1 without any increase in exhaust-gas temperature. Extrapolating these curves, it appears output will continue to increase, probably up to a pressure ratio of 10:1.

What this points out is a correctly turbocharged and intercooled diesel engine will continue to increase in power at manifold pressures beyond the capabilities of today's engines.

Büchi also pointed out that higher

Doug Drussel of Garden City, Kansas, set up this John Deere 4010 with three-stage turbocharging. Water is injected at a rate of three gallons per minute for intercooling. Manifold pressure is 175 psig at 4000 rpm.

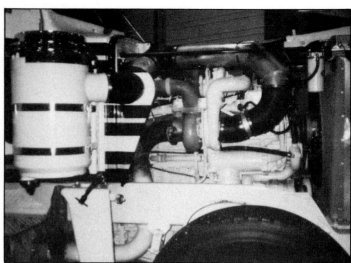

People laughed at the first pulling tractors with staged turbochargers, but this production Cummins in a Peterbuilt is not a joke. It produces 475 HP all day at 2100 rpm. The first stage is AiResearch and the second is Holset.

outputs do not necessarily increase combustion temperatures. Dr. Büchi was running 71-psig intake-manifold pressure in 1909! He probably would have run higher pressures if materials that could stand the added stress had been available.

As a result of high-pressure turbocharging with intercooling, tractor pullers are obtaining up to 800 HP at 4000 rpm from a 400-CID diesel. This is from an engine designed to produce approximately 175 brake horsepower at 2500 rpm.

MATCHING STAGED TURBOCHARGERS

If each compressor produces a 2.7:1 pressure ratio, the interstage absolute pressure will be 2.7 x 15 = 40.5 psia. Intake-manifold pressure will be 2.7 x 40.5 = 109 psia, or 94 psig. Some tractor pullers have reported intake-manifold pressures as high as 130 psig using this method.

When two turbochargers are used in tandem, the output of the first-stage compressor must be matched to the input of the second-stage. Matching must also account for any intercooling between stages. This is done "backward," just as a building is designed from the top floor, not the basement.

Second Stage—As an example, a 400-CID engine running at 4000 rpm with 80% volumetric efficiency will use 375 cfm, Figure 4-1. This means a

compressor capable of producing 375 cfm at a 2.7 pressure ratio must be used for the second stage. Assuming 65% compressor efficiency and 2.7 pressure ratio, the density ratio is 1.8.

This means the second-stage turbocharger must flow:

1.8 x 375 = 675 cfm

Either an AirResearch T04B V-1 Trim, Rajay 370E or a Schwitzer 4LE-354 will handle this.

Intercooler—Choosing the first-stage compressor is a little harder because of the intercooler between the two compressors.

Using Don Hubbard's Chart, Figure 16-3, at 2.7 pressure ratio and 65% compressor efficiency, discharge temperature of the first stage is 360F. Ice water is used as the cooling medium, so intercooler temperature is 32F.

An intercooler effectiveness of 75% will reduce the temperature 75% of the difference between the two temperatures. Subtracting 32F from 360F, the difference between the compressor-discharge temperature and the cooling medium is 328F. Multiplying this by 0.75, the temperature drop across the intercooler is 246F.

A 246F reduction means the temperature of the air entering the second-stage compressor is 360 − 246 = 114F. This means 675 cfm of

air will enter the second-stage compressor at 114F.

First Stage—Assuming a negligible pressure drop through the heat exchanger, the flow from the first-stage compressor is:

$$CFM_2 \times \frac{T_1}{T_2} = 675 \times \frac{360 + 460}{114 + 460}$$

$$= 675 \times \frac{820}{574}$$

$$= 964 \, CFM_1.$$

where CFM_2 is the flow into the second-stage compressor and CFM_1 is the flow from the first-stage compressor into the intercooler. The difference between the two is caused by the intercooler. All ratio calculations must be done in degrees Rankine, so 460 must be added to each of the temperatures.

Output from the first-stage compressor is 964 cfm. It's best to keep the pressure ratios of the two compressors about the same, or 2.7 in this instance. Consequently, the density ratio will also be the same, 1.8.

Multiplying 964 x 1.8 = 1735 cfm gives the required capacity of the first-stage compressor. This is where things get "hairy." We are now talking about a large turbocharger.

The horsepower obtained from these engines is still a function of the amount of air pumped through them. The first-stage turbocharger must be capable of supplying air to an 800-

horsepower engine. The first-stage compressor doesn't know it is blowing the air into another compressor so the air can be compacted to a point where it can fit into a 400-CID engine.

AiResearch T-18 turbochargers have been used successfully. There are many versions of this turbo, so be sure to pick the right one. Model T18A40 with a 1.50 A/R turbine housing has worked well.

A little luck will be required in picking the correct turbine-housing sizes the first time. You'll need two pressure gages—one on the intake manifold and one between the stages. With the gages, you can determine if each turbocharger is running at the same pressure ratio. Remember, in each case it is necessary to add atmospheric pressure to the gage pressure when calculating pressure ratios.

Some engine builders use an after-cooler between the second-stage and the engine in addition to the intercooler between stages. The aftercooler means you'll have to calculate the increase in flow through the second-stage compressor. This is done the same way as it was for the first stage.

THE NEXT STEP

Now that you've digested the above, I must tell you the method is obsolete. Manifold pressures of 100 psi are not good enough to win consistently. The logical—are you kidding?—next step was to add a third-stage turbocharger. With three stages of turbocharging and three intercoolers, the engine disappears under a maze of ducting.

There is, however, an alternate method to cool the charge. It wasn't long before some brave puller tried it. It consists of spraying water into the discharge of each stage of compression. This eliminates the bulk of the heat exchangers (intercoolers). Assuming we still have 2.7 pressure ratio per stage, we end up with an intake-manifold pressure of:

14.7 x 2.7 x 2.7 x 2.7 = 289 psia or
= 274 psig

This works out to somewhere around 1800 HP for a 400-CID diesel engine.

I have not attempted to do the turbocharger matching on such an engine, but the fact that it works at all is amazing. We considered two-stage turbocharging as exotic only a few years ago. Now Cummins has such an engine in production.

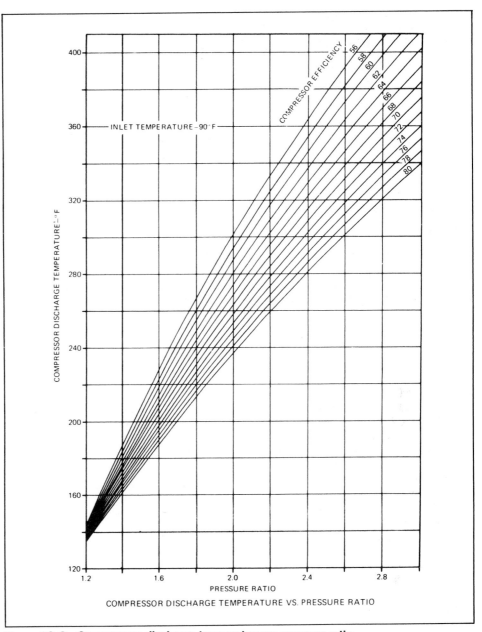

Figure 16-3—Compressor-discharge temperature vs. pressure ratio

The engine builders who specialize in tractor pulling are truly pioneers. The things they are doing today may be commonplace tomorrow.

Dyed-in-the-wool drag-race fans may laugh at the thought of a tractor-pulling contest, but there are many similarities between them. The super-stock class is the tractor version of the funny car. The pulling contest, however, has one big advantage. You can cheer your hero on for two minutes instead of six seconds. And, you can see the participants from start to finish!

17 Maintenance

Porsche turbocharged and intercooled Indy Car engine was never raced because of rules problems. Note blowoff valve at end of plenum chamber. Although your turbocharged engine may not be for racing, it requires the same basics: clean oil and air. Photo courtesy of Interscope-Porsche.

Now that turbochargers have come of age, users are not quite as ready to blame engine failures on them as they were a few years ago. Still, far too many turbocharger and engine failures occur from lack of maintenance. Maintenance takes only a little time and can save many dollars and heartaches.

I've looked at enough failed turbocharger parts to be able to tell if the failure was due to manufacturing problems or some other cause. In the vast majority of cases, the failures were caused by lack of maintenance or misuse.

Using the information in this chapter during installation and operation could prevent them.

Chapter 2 describes the design and operation of a turbocharger. Chapter 4 covers sizing and matching the turbocharger to the engine. Assuming both were done correctly, the turbocharger will run at least as long as the engine, unless it is mistreated. The next few pages cover typical turbocharger failures and why they occur.

JOURNAL/BEARING FAILURES

These failures are about as common as any other type.

Oil Starvation—Journal/bearing failures are caused frequently by lack of lubrication, which is easy to spot. The shaft journals will be discolored from the heat generated at high speed with no lubrication, Figure 17-1.

Some discoloration may occur from varnish buildup on the turbine-end journal after many hot shutdowns. Discoloration from lack of oil will appear straw colored or deep blue. The color is almost the same as would occur from tempering the part after heat treating.

Journal/bearing failures could be caused by a broken oil-supply line, lack of oil in the engine or too-heavy oil viscosity for the ambient-temperature conditions.

Bearing failures can also be caused by placing an orifice in the oil-supply line to reduce oil pressure to the turbocharger. One speck of dirt can clog a small orifice and cause failure in a matter of seconds.

Dirty Oil—Journal/bearing failure will also occur when dirt gets into the oil. This type of failure, Figure 17-2, will not necessarily result in discolored journals. Instead, there are usually grooves in the journals from dirt em-

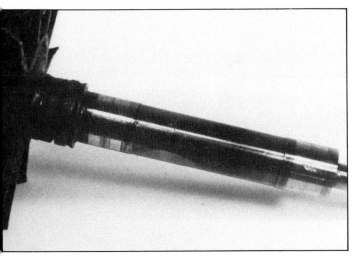

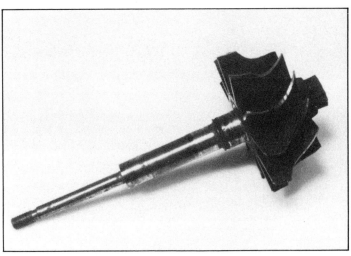

Figure 17-1—Journal and bearing failure caused by lack of lubrication. Dark areas are actually straw to dark blue in color. Bearing material is welded to journal. Photo by Jim Miller.

Figure 17-2—Journal and bearing failure caused by dirty oil. Both journals have been grooved by particles in oil. Photo by Jim Miller.

bedding in the bearings.

Another symptom of this type of failure is bearing material welded to the shaft journals. This could happen for one of several reasons.

Dirty working conditions during installation can allow dirt into the turbocharger and cause this kind of failure. Running an engine without an oil filter is another cause. Not changing a bypass-type oil filter will allow dirty oil to flow around the filter element into the turbocharger journals.

There has been much controversy as to whether an oil filter should be installed in the turbocharger oil-inlet line. The problem is that if the filter clogs it can cause a failure by restricting oil flow to the turbocharger. As noted earlier, failures also occur when a bypass filter clogs, allowing dirty oil to flow through the bypass.

Both of these failures can be prevented by changing the oil filter at regular frequent intervals. This is one place where it certainly is not economical to save a couple of dollars. As the man used to say in the TV oil-filter ads, "Pay me now or pay me later."

TURBINE & IMPELLER DAMAGE

Anyone who builds a racing engine is usually careful to cover all the inlets to the carburetor or fuel-injection system when the engine is not running. Dirt and foreign objects are very destructive to any kind of engine. Despite this, one of the major causes of turbocharger failure is a foreign object tangling with the compressor impeller, Figure 17-3.

Foreign Objects—If a nut or bolt enters the compressor when it's running at high speed, it's like putting something in a meat grinder. The problem is that, in this case, the meat is tougher than the grinder. To make things worse, pieces chewed off the impeller will enter the engine along with the object that started the damage.

This type of failure is about as bad as they come, because it may mean a new engine as well as a new turbocharger. For this reason, always use an air cleaner on any engine, particularly a turbocharged one. If the correct filter is used, there should be no horsepower loss.

When building a naturally aspirated engine, little concern is shown for dirt in the exhaust manifold. Nuts or bolts left there will blow out with no harm done. Not so with a turbocharged engine.

Several years ago, it was established that a turbocharger is an excellent spark arrester. These are required for engines operating in areas where there is a danger of fire, such as dry forests. Any particles much heavier than air hit the tips of the turbine blades and are knocked back into the turbine housing. Once the particles come out they are so small they are no longer considered a spark.

The same thing happens with a nut, bolt or a small piece of exhaust valve or piston ring. Whenever something gets into the exhaust system of a turbocharged engine, it hits against the turbine blades until the tips look like a mouse has been chewing on them,

Figure 17-3—Compressor-impeller damage caused by foreign object. Photo by Jim Miller.

Warner-Ishi turbocharger undergoing inspection after racing 500 miles at Pocono—it was on Rick Mears' winning Indy Car. Shaft end play is being checked here. Black streaks are from debris—mostly rubber particles—drawn into compressor with unfiltered air.

121

Figure 17-4—A piece of the engine caused this turbine-wheel failure. Photo by Jim Miller.

Figure 17-5—Compressor-impeller hub burst from overspeeding. This type failure usually results in the impeller passing completely through compressor housing. Photo by Jim Miller.

Figure 17-6—Turbine-wheel failure caused by excess temperature and/or overspeeding. This is a minor failure because only one blade broke off. When the hub bursts, it is much more damaging—and dangerous. Photo by Jim Miller.

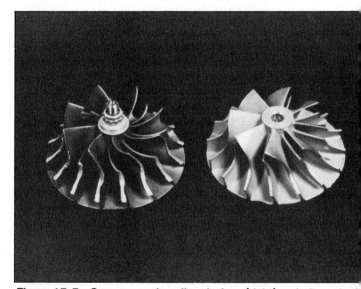

Figure 17-7—Compressor impellers before (right) and after a bad backfire (left)

Figure 17-4.

This can occur without loss of balance of the turbine, shaft and impeller. The turbocharger may run for thousands of hours without further damage. Most of the time, however, one of the blades will break off and the imbalance causes immediate bearing failure.

Overspeed—Some turbocharger users brag about getting 50 psi out of a compressor when the map for the unit only goes to 30 psi. They may be lucky for a while. Eventually they'll end up with a turbine or compressor wheel looking something like Figure 17-5 or 17-6.

When turbine-wheel or compressor-impleller failure occurs from overspeeding, it is like a hand grenade going off in the turbine or compressor housing. The higher the speed, the more dangerous it is.

Backfires—A backfire through the compressor—not a little cough on a cold morning, but a big sneeze—at or near maximum speed and power will cause compressor-impeller damage.

If a compressor impeller looks like the one in Figure 17-7, it probably got that way by trying to reverse direction while rotating about 120,000 rpm. This type of failure is often accompanied by a broken shaft.

Obviously, the way to prevent this is not to let the engine backfire. Backfires are usually caused by ignition—more properly lack of ignition.

If a cylinder fails to fire, the charge will be exposed to the hot exhaust gases at the beginning of the exhaust stroke. This can ignite the charge. When the intake valve opens, all of the charge in the intake manifold may ignite.

Turbocharger manufacturers are reluctant to honor a warranty claim resulting from an obvious backfire. The problem is with the ignition or fuel system, not the turbo. Port fuel injection minimizes the amount of fuel in the intake manifold and should reduce this type of failure.

OIL LEAKS

The most frequent complaint about turbochargers is they leak oil. Although it's always blamed on the turbocharger, oil leaks are usually an installation or engine problem.

Many turbochargers use a simple labyrinth seal at the turbine end. This type of seal does not leak oil any more or less than those with a piston-ring-type seal.

Improper Oil Drains—As mentioned before, oil enters the turbocharger from the engine system *at engine pressure*. After the oil passes through the bearing it must flow *by gravity* out of the bearing housing and back to the engine.

Anything restricting the oil-drain line will cause the oil level in the bearing housing to rise above the oil seals. With a labyrinth or piston-ring seal, the oil will leak into the end housings. This is probably the main cause of oil "leakage," although other things will contribute as well.

Engine Vacuum—At the compressor end, if the air cleaner is allowed to get very dirty it creates a sizable pressure drop through it. This can create a slight vacuum behind the compressor impeller. This is particularly true in a blow-through application at idle.

If the turbocharger has a piston-ring seal on the compressor end, oil can be sucked into the compressor housing. Lots of blue exhaust smoke results.

Piston-ring-type seals can handle only a very small vacuum, so this type of seal is not recommended when sucking through the carburetor. Vacuum as high as 29 in.Hg abs. will be imposed on the compressor.

Varnish Buildup—Rajay, Roto-Master and certain models of AiResearch T04 Turbochargers use a mechanical face-type seal. If it leaks, it's probably gummed up with varnish and is not able to float freely. If the seal leaks it should be cleaned or re-

placed immediately. Otherwise dirt will enter the space between the carbon-seal face and mating ring, causing rapid seal failure.

Other oil-drainage problems which can frequently cause oil leakage are detailed in Chapter 9 on Lubrication. The table on this page shows a list of problems that occur on turbocharged

engines and the probable causes. Whenever possible, try to find a remedy that does not include disassembly of the turbocharger because *it probably is not the culprit.*

A turbocharger can last as long as the engine if it is installed correctly, supplied with clean oil and prevented from swallowing solid objects.

TABLE 9
CARE AND MAINTENANCE
Typical problems causing turbocharged-engine malfunction

Symptom	Cause	How to Check	Remedy
Lack of boost	Gasket leak or hole in exhaust system	Block off tailpipe with engine running. If engine continues running, leaks are present	Repair leaks, usually gasket surfaces
Lack of boost	Worn valves or rings	Compression check engine	Repair
Lack of boost	Carburetor too small or butterfly does not open completely	Check pressure drop through carburetor	Use larger carburetor or adjust linkage
Lack of boost	Restriction in turbine discharge system (muffler)	Check turbine discharge pressure	Use larger muffler or larger diameter pipes
Lack of boost	Dirty air cleaner	Remove air cleaner	Service air cleaner
Gasoline odor during boost conditions	Small leak at compressor discharge intake manifold	Look for fuel stains around joints	Tighten joints or replace gaskets
Poor throttle response (stumble)	Clogged circuit in carburetor	Try another carburetor or richen jet(s)	Clean carburetor and check jet sizes
Plugs miss at high power	Gap too large	Measure gap	Clean and reset to 0.025 in.**
Plugs miss often*	Bad leads	Check lead resistance against specifications	Replace leads
Oil leak into turbine housing	Blocked oil drain	Remove drain line and check for plugged or crimped line	Clean or replace drain line
Poor idling	Air leak between carburetor and compressor	Listen for hissing around carburetor at engine idle	Repair leak

* At boost of 15 pounds or more, use the best capacitor-discharge ignition system available, preferably with a high-output coil.

** Smaller gaps may be required if misfiring continues.

Rabbit/Scirocco is fitted with sanitary turbocharger kit. Callaway also supplies intercooler. When looking for a kit for your engine, make sure it will fit and that all hardware is included. Photo by C. E. Green.

Bob McClure at B.A.E. installed twin turbos on his Ferrari Boxer. Exceptionally neat installation uses B.A.E.-adapted fuel injection. Performance is as good as appearance.

The number of production passenger cars with factory-installed turbochargers has exploded from a single one in 1975. Virtually every major auto manufacturer offers one or more turbocharged models.

During this period, the small spark of interest in turbocharging cars in the aftermarket has grown into a raging fire. I have tried to list as many makers and installers as space permits, but many new ones are sure to appear after this revision is printed.

The number outside of the United States has grown even faster than inside the United States. In this chapter, I have included additional information about some representative companies.

A few years ago, passenger-car turbocharger kits—particularly bolt-on—were such a novelty that automobile magazines were glad to give them space. Today there are so many kits, it's hard to tell whose kit is on an engine unless a nameplate is hung on it.

This is also true for motorcycles. Motorcycle kits are discussed separately in Chapter 21. Bolt-on kits and complete marine engines are covered in Chapter 11. Some companies offering kits are listed in the back of this book.

AUTOMARINE TURBOCHARGERS, CHRISTCHURCH, ENGLAND

Malcolm Cole, a graduate of the Janspeed School, is not new to turbocharging. Cole likes the challenge of things that are not run-of-the-mill. He produces a kit for the Mazda RX7 Rotary that cuts two seconds off the 0-60 mph time and 11.3 seconds off the 0-100 mph time.

B.A.E., TORRANCE, CALIFORNIA

B.A.E. started in 1968 as Bob's Automotive in Lawndale, California. Beginning with customer installations of turbochargers on street and race cars and on marine engines, Bob McClure directed his efforts toward the creation of kits for marine and automotive applications.

Gale Banks Engineering twin-turbo kit for small-block Chevy includes exhaust manifolds with mounting pads for dual waste gates. Version of kit is shown on cover.

GIK kit for Volvo 242, 244 and 245 models used Roto-Master turbocharger and waste gate.

In 1977 the name was changed to B.A.E. and the firm moved to larger quarters. Former Rajay engineer Bill Wilbur joined the firm to assist in engineering.

By 1983 the firm offered over 50 separate kits for U.S. and foreign cars, trucks, motor homes and vans. These include BMW 4- and 6-cylinder engines; Capri 2.6 and 2.8; Chevrolet small-block for cars, pickups and Blazers; Datsun 510, 520, 240Z, 260Z and 280Z; Dodge 440-CID motor home; Fiat X1/9; Ford trucks 360, 390 and 460 CID; Honda Civic, International diesel (Nissan); Mercedes diesels 220D, 240D and 300D; Porsche 914, 924 and 911 with CIS; Saab; Toyota T2-C, 20 R.C.; VW air-cooled Types I and II, Rabbit and Scirocco carbureted or injected.

B.A.E. has California Air Resources Board (CARB) approvals for kits on the Dodge 440, Volkswagen Rabbit (fuel-injected and diesel), Isuzu diesel, and BMW 320i.

GALE BANKS ENGINEERING, S. SAN GABRIEL, CALIFORNIA

Gale Banks is noted for his many successful applications of turbochargers to marine engines. He still likes to apply his skills to passenger cars, even if it's only for the satisfaction of seeing a Corvette go over 200 mph at Bonneville. Gale's twin turbocharged blow-through small-block Chevy is shown on the cover. It puts out 580

HP on 92-octane gas and still fits under the hood.

Gale Banks is discussed in the chapter on marine engines. However, the company also offers turbos for passenger and race cars, primarily for U.S.-made engines.

CALLAWAY TURBOCHARGER SYSTEMS, OLD LYME, CONNECTICUT

Reeves Callaway is a perfectionist who insists his kits must look professional in addition to running well. His fuel-enriching device can be installed on either carbureted or injected engines. It allows the fuel system to be set for good economy without worrying about running too lean when the turbocharger starts creating boost. Callaway also offers intercooler kits for many factory-turbo installations.

GIK TURBOTEKNIK, GOTEBORG, SWEDEN

This company, run by Erland Myhrman, has been making kits since 1976. They have bolt-on kits for Mercedes 220D and 240D, Mercedes 300D, Volvo B-20 and B-21 engines. In addition, they have Workshop Kits for about 20 other engines. The Workshop Kits come with all necessary material, but welding is required to install one.

Kit builders in Sweden are faced with emission rules similar to those in California.

HKS TURBO, SHIZUOKA, JAPAN

HKS kits are popular in Japan and are sold through most speed shops. In addition to Roto-Master, they use AiResearch and Hitachi Turbochargers.

HKS not only develops, manufactures and markets kits, but they sell low-compression pistons and complete bolt-on exhaust systems as well. These are truly complete kits, so the customer does not have to fabricate anything.

JANSPEED ENGINEERING, SALISBURY, WILTSHIRE, ENGLAND

Jan Odor came to England from Hungary in 1955. He proved it is possible to build the number-one turbocharger kit business in Europe from nothing, if one has lots of energy and know-how.

Before Jan became interested in turbocharging, he raced Datsuns and manufactured exhaust headers. He demonstrates his products personally, and a test drive with him is an exhilarating experience.

What Jan did not know about turbocharging he learned by experimenting. Janspeed now makes kits for just about everything available in Europe.

Janspeed has kits for blow-through and draw-through carbureted applications, fuel-injection and diesels, all

HKS kit for Toyota Supra

Janspeed installation fits either Honda Prelude or Accord.

M & W uses water-jacketed turbine housings on all of their marine installations.

Ak Miller's open-wheel Pikes Peak entry for 1977 used dual AiResearch TO4 turbochargers, Impco propane carburetion and boost-limiting valves. Engine is 351C Ford.

using Roto-Master turbochargers. His kits are notable for being complete and attractive as well as giving outstanding performance.

M & W GEAR, GIBSON CITY, ILLINOIS

M & W is the most successful kit manufacturer in the United States, if not the world. Elmo Meiners and Jack Bradford started making turbocharger kits for farm tractors back in 1962. Within a few years they had turbochargers on just about everything that pulled a plow or picked a bean.

When turbocharging became factory equipment on most farm tractors, they started making kits for marine engines, both gasoline and diesel. They now have kits for most of the in-line and V-type marine engines, all of which include water-jacketed turbine housings.

All M & W kits are thoroughly tested in the field or on the water before being released for production.

AK MILLER ENTERPRISES, PICO RIVERA, CALIFORNIA

Ak Miller and his cohorts Jack Lufkin, Burke LeSage and Bill Edwards have been in the hot-rod business since before they were called *hot rods*. These fellows, along with Duke Hallock of AiResearch, have been using turbochargers to get added horsepower since the '50s.

It is only natural that Ak should offer turbocharger kits. In some cases he has combined them with propane

fuel for the "cleanest" turbocharger kits anywhere.

Ak has kits for the 2-liter Ford (Pinto), Ford-6 170, 200, 240, 250 and 300 CID, Chevy 6, Capri V6, fuel-injected Porsche 911 and Datsun 280Z. Truck kits are designed for Chevrolet, Dodge and Ford pickups, as well as for the Toyota and Courier mini-trucks. There is also a kit for the Ford 534-CID industrial engine.

Ak handles Roto-Master turbochargers. Other items, such as car-

Primary exhaust pipes join at collector just before turbine. Spearco installation uses Warner-Ishi RHO5 turbocharger on VW 1600 Type 1 air-cooled engine.

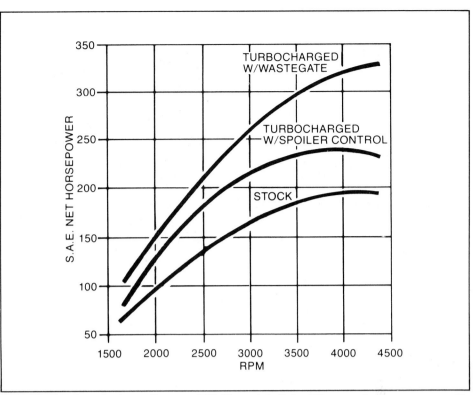

Roto-Master dyno-test results on a 350 Chevrolet. Note difference in power curve with spoiler-type boost control and waste gate boost control.

buretor adapters and manifolds, are available. He also sells the IMPCO flow-control valve, which prevents the engine from being overboosted.

Ak has done considerable testing in the emissions area, showing that turbocharging an engine can improve overall emissions.

SPEARCO PERFORMANCE PRODUCTS, INC., INGLEWOOD, CALIFORNIA

Besides turbochargers and kits, George Spears offers electronically controlled water injection, electronically controlled spark retard, waste gates, instruments, and other bits and pieces for installers and kit builders.

TURBO CITY ORANGE, CALIFORNIA

Turbo City specializes in mail order and over-the-counter sales for the guy who wants to build his own system. They offer the full Roto-Master (and Rajay) line of automotive turbochargers and wastegates. They also carry a full line of adapters, manifolds, and accessories for custom installations.

TURBO PERFORMANCE PRODUCTS, N. HOLLYWOOD, CALIFORNIA

Roy Findlon has been associated with turbochargers since the Corvair days. He still does performance work

Installation by Turbo Performance Products fits neatly into Porsche 912 engine compartment. Boost is controlled by IMPCO valve. Note engine-oil cooler attached to engine cover.

on them, but has branched out and includes other engines as well.

Roy is known for neat work and his shop always looks like he is ready for a white-glove inspection. This cleanliness is reflected in his installations, which run as good as they look.

TURBO SYSTEMS, INC., AKRON, OHIO

Bill Laughlin of Turbo Systems decided a few years ago that turbochargers were the only way to go. He purchased the Pinto and Vega turbocharger kit lines from Car Corp. of

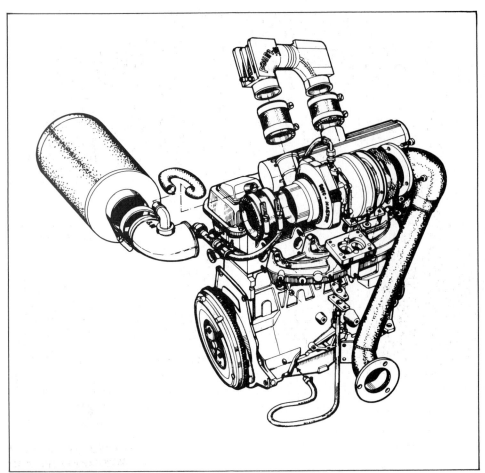

Hi-Tuned kit for Rabbit/Golf diesel

If a turbocharger kit or turboed production car doesn't include an intercooler, an intercooler kit may be available. This HKS intercooler is used on the Toyota Supra Turbo. Photo courtesy of Luckey Dodge.

Livonia, Michigan, and has since added a small-block Chevy kit of his own.

The small-block kit is a real performance kit using two turbochargers. This kit will fit 265—400-CID small-block Chevrolets. Bill has cast new exhaust manifolds for the engine, so the turbochargers can be mounted high on the front of the engine. This kit is for power rather than comfort—air conditioning cannot be used.

Turbo Systems has also done a lot of work with diesel tractors used in tractor-pulling contests.

TURBOCHARGER, INC., DOWNEY, CALIFORNIA

Many diesel kits have been offered for trucks, farm tractors and construction, but this one is for a passenger car. It is no secret that the Mercedes diesel-powered cars are not very exciting when it comes to acceleration.

This lack of performance is not due to the fact it is a diesel but rather, it is a *small* diesel. The largest four has a displacement of only 2.4 liters or 144 cubic inches.

Turbocharging a diesel has all of the advantages of turbocharging a gasoline engine plus a few more. These are: improved fuel economy, smoke reduction and lower emissions.

The performance of this kit as measured by *Road and Track* (September 1973) improved acceleration markedly. The 0—60-mph time was reduced by almost 10 seconds—not bad by any standards.

The kit fits all series 200, 220 and 240 diesels with either manual or automatic transmissions.

MR. TURBOCHARGER, ELANDSFONTEIN, SOUTH AFRICA

Working with Associated Diesel Company of Johannesburg, Mr. Turbocharger has developed kits for the following engines:

Datsun—Pulsar 1800, 2000, 2800 Fuel Injection

Dodge Colt—1600, 2000 and 2600

Ford—1600 Sport, 2000 OHC and 3000 V6

Volkswagen—Rabbit/Golf Diesel and Petrol, 1500 and 1600 CC

Isuzu—KB Diesel

Mazda 323—1300, 1400 and 1600

Mercedes Benz—240D and 300D

Opel Kadett—1300

Toyota—1600 and 1800, Landcruiser Diesel and 4 x 4 Hi-Lux

Ford and Chevy V8—all sizes

This might change your mind about Africa being all jungle and desert.

VAN ZIJVERDEN TURBO B.V., HOOFDDORP, HOLLAND

Looking at Van Zijverden's Hi-Tuned Kits, especially the Mercedes 240D, the first reaction is usually "Why can't *we* make one as neat and simple?"

There is one reason. When a kit is developed in the U.S., it must fit around the air-conditioning system. Few cars in northern Europe are air conditioned, so the kit builders can pay more attention to aerodynamics than to available space.

There are Hi-Tuned Kits for the VW Rabbit/Golf, both gasoline and diesel, Mercedes 200D and 240D, Mercedes 300D, Ford Granada 2.1D, Mitsubishi Colt and Alfa Romeo.

19 Exhaust Emissions

One of the questions asked frequently about turbochargers is, "What will a turbocharger do to exhaust emissions?" There is no simple answer. A turbocharger is not a device to chop exhaust emissions into little bits and spit them out as clean air.

However, tests have shown that a turbocharger on a diesel or a gasoline engine in good running condition will not significantly increase any measured exhaust emissions. In fact, a turbocharger will decrease them in almost every case. The nice part is that it can be done while increasing the output of the engine by as much as 100%.

When equipped to meet U.S. exhaust-emissions requirements, the standard gasoline engine loses a good deal of power. Vehicles no longer perform the way many of us have accepted as standard.

A turbocharger will allow a gasoline engine to produce the horsepower it had before emissions devices were installed. The major disadvantage here is fuel consumption. It will still not be as good as a high-compression, naturally aspirated engine.

With the diesel engine, it is a slightly different story. Until recently, diesel engines were not seriously considered for U.S. passenger cars. The reasons included a poor power-to-weight ratio and high initial cost.

Even after all the emissions-control devices were added to gasoline engines, few people took the diesel engine seriously. The extra fuel burned in the gasoline engine wasn't all that expensive—at least in the U.S.

Events of the past few years have altered our thinking considerably. In some countries it's considered patriotic to try to reduce fuel consumption as well as exhaust emissions.

A diesel like the Nissan CN6-33 198-CID in-line six will run 4000 rpm—fast enough for any passenger car or pickup truck. Naturally aspirated, this engine has a maximum output of 92 HP at 4000 rpm.

In weight and power this engine is almost competitive with a current small, naturally aspirated gasoline V8.

Part of B.A.E. emissions lab illustrates equipment needed to monitor carbon monoxide (CO), hydrocarbons (HC) and oxides of nitrogen (NO$_x$) during emissions testing. Once approved, recertification of a specific kit is difficult. This is due more to paperwork than engineering! Approval is granted only for specific years of an engine. Any changes to kit or engine requires recertification.

Northrop Institute of Technology entry in 1973 and 1974 Reduced Emission Rallies for college students was a turbocharged Toyota using propane fuel. One-third of cars entered in 1974 had turbocharged engines.

I say "almost" because a free-floating turbocharger makes it more than competitive. The turbocharged version produces approximately 130 HP.

The state of California started conducting tests with this engine in January 1974. In the tests, fuel consumption was considered as important as

TABLE 19-1
**Fuel-consumption & exhaust-emissions comparison of a
1973 3/4-ton truck with gasoline & diesel engines, both air-conditioned**

COLD START CVS TEST

Truck	Engine	GRAMS/MILE HC	CO	NOx	Miles/ Gallon
CHC 2747	Base line gasoline	3.72	53.82	4.93	11
CHC 2748	Base line gasoline	2.91	37.84	4.52	11
CHC 2747	Base line diesel	1.17	4.14	2.52	20
CHC 2748	Turbocharged diesel	0.47	2.54	2.13	20

ADDITIONAL HOT CVS—1 TEST

		HC	CO	NOx	
CHC 2748	Emissions	0.26	2.10	1.85	

	SPEED (MPH) 30	40	50	60
Dynamometer Fuel Consumption	29.4	26.2	23.8	20.3

TABLE 19-2
COMPRESSION PRESSURE OF VARIOUS ENGINES CALCULATED FROM FIGURE 19-1

	Naturally Aspirated		Turbocharged	
Compression ratio	10.5/1		7.0/1	7.0/1
Intake-manifold pressure psia	14.7	P_b	29.4	25
Inducted-charge temperature R°	600	T_b	750	700
Inducted-charge density ft³/Lb.	15.05	V_b	9.41	10.5
Entropy	0.085	S	0.095	0.09
Charge density at end of compression stroke—ft³/lb.	1.43	V_c	1.345	1.53
Compression pressure psia	340		359	320

low exhaust emissions.

Tony Capanna of Wilcap made the installation for the State of California. He had been running several diesel-powered passenger cars.

One of Capanna's cars is a Dodge Dart with the same turbocharged diesel as in the state pickups. Because of its light weight, the Dart gets 40 miles to the gallon. Tony doesn't have to worry about towns that don't have diesel fuel—he might drive through an entire state without refueling.

The original test was run on two Dodge 3/4-ton pickups normally equipped with 360-CID gasoline engines. One of the trucks had a naturally aspirated CN6-33 Nissan engine. The other has the same diesel engine with a Rajay turbocharger.

Both trucks were tested for exhaust emissions and fuel consumption before the gasoline engines were removed. Both tests were rerun after the diesel engines were installed. Comparative results are shown in Table 19-1.

As a baseline, the results were compared to the then-proposed 1976 California requirements. The turbocharged diesel was twice as good on HC, seven times as good on CO and only 7% too high on NOx. Perhaps even more interesting, emissions from the turbocharged diesel were lower than those from the naturally aspirated diesel.

The nicest thing about the results was that the engine used no special devices that increased fuel consumption without producing power. When looking under the hood of a gasoline-powered emissions-controlled car, Patricia Spears once made the comment, "The only people who make out from all these emissions controls are the hose manufacturers."

LEGAL REQUIREMENTS

Many kit manufacturers and installers have been concerned about the possibility of breaking the law by adding a turbocharger to an engine. After applying to the California Air Resources Board for an exemption from Section 27156, several manufactures discovered there was no set procedure or precedent for exemption.

Aftermarket Certification—Roto-Master, Inc. decided to pursue an exemption regardless of the amount of time and effort required. The people at the company wanted see if it could be done.

The vehicle chosen was a 1976 Mercedes 240 diesel. The kit was developed, tested and submitted to CARB for exemption.

One of the requirements for certification is that the turbocharged vehicle can have no more emissions than the naturally aspirated version, *even if the modified version is still within the limits set by law.* This means that if your baseline vehicle is particularly good naturally aspirated, you're in trouble.

Roto-Master presented complete installation information and operating schematics of the system. A low-mileage, naturally aspirated car was tested by CARB for baseline data. The kit was then installed on the vehicle, by CARB, and the comparison emissions tests were run. The vehicle was also tested for driveability.

The turbocharger was removed from the vehicle and it was rechecked naturally aspirated. The turbocharger was reinstalled and all tests repeated.

These tests proved only that the turbocharger installed on a low-mileage car did not increase emissions, despite a substantial increase in power. However, it did not prove the turbocharged car would conform over thousands of miles under various conditions.

Fortunately, Roto-Master had nine other 240Ds undergoing tests: four in California, one in New Mexico, two in Germany and one each in England and Finland.

Some of these cars had accumulated more than 50,000 miles during the course of the test. The combined experience of all the test vehicles convinced CARB to grant an exemption for this kit. It was Executive Order

D-89 of April 24, 1979.

Unfortunately, the whole procedure took so long that by the time the kit was certified, Mercedes had redesigned the car and the engine. The new car uses a completely different fuel-injection system, which made the turbocharger system obsolete. Roto-Master decided not to produce the kit.

They say some good comes from everything. This exemption broke the ice for other kit manufacturers by setting a precedent and establishing a procedure.

Several kit manufacturers have received exemptions since then, including RV Turbo, Inc., D-90, on September 20, 1979, for the Dodge 440 engine in a motor home. This, subsequently, was reissued as D-90-1 to Turbo International, which produces a similar installation.

On February 29, 1980, Bob Keller obtained Executive Order D-99, covering exemption of a kit for a Chrysler 360-3. This is also a heavy-duty engine used in a motor home. Several others have been issued or are pending. It can be done!

Now that it's been established that a turbocharger will not hurt and will usually improve exhaust emissions, it's only fair to ask, "Why?"

Diesel Emissions—In the case of the diesel engine, the answer is relatively simple. Air flow through a naturally aspirated diesel is strictly a function of engine speed. Power demands have little effect on air flow, because a diesel does not normally have a butterfly in the air intake.

Output is governed by the amount of fuel injected with each power stroke. Maximum output of a naturally aspirated diesel is limited by exhaust smoke.

A turbocharger developing manifold pressure of 10 psig—which is easy to obtain—will increase the air flow through a diesel about 67%. If the fuel setting is left the same as when naturally aspirated. The extra air lowers combustion temperatures. And, lower combustion temperatures lowers NO_x emissions.

The extra air will also produce more complete combustion. The additional oxygen present will reduce hydrocarbons and carbon monoxide.

Diesel power output is increased by burning more fuel per stroke. However, emissions will be reduced as long as the fuel/air ratio is kept considerably leaner than when naturally aspirated.

Gasoline-Engine Emissions—The gasoline engine is a different story. Here the fuel/air ratio must be close to a stoichiometric mixture or the mixture will not burn.

This means the fuel/air ratio of the turbocharged engine will have to be close to the same as on the naturally aspirated engine. Any improvement in the exhaust emissions after the turbocharger is installed is due to some cause other than a leaner fuel/air ratio.

The key lies in engine design. No naturally aspirated gasoline engine has a perfect induction system. Consequently, the fuel/air ratio will vary somewhat from cylinder to cylinder. Some of the cylinders will operate lean while others will run slightly rich.

Those that are slightly lean will tend to increase the output of NO_x. Those that are slightly rich will tend to increase HC and CO_2.

When the carburetor is upstream of the compressor, all the fuel passes through the compressor impeller on its way to the combustion chamber. The compressor helps improve fuel vaporization and distribution.

Figure 19-1—Compression chart for chemically correct octane/air mixture plus average amount of clearance gases.
Line A is naturally aspirated, 10.5:1 compression ratio.
Line B is supercharged at 2:1 pressure ratio, 7:1 compression ratio.
Line C is supercharged at 1.7:1 pressure ratio, 7:1 compression ratio.
Lb fuel/lb air = 0.0665; E of combustion = 1280 (1-f); H of combustion = 1278 (1-f).
(From Hershey, Eberhardt and Hottel, Transactions of the S.A.E., October 1936).

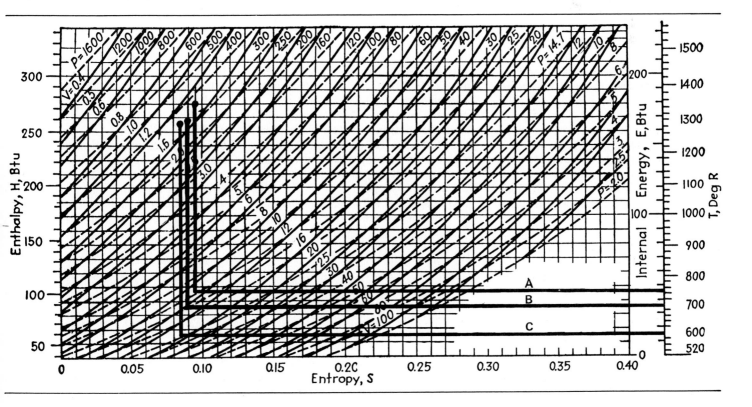

TABLE 19-3

TURBOCHARGED TCCS ENGINE-POWERED M-151 VEHICLE

Emissions and fuel economy with gasoline fuel & without emissions controls

| | EMISSIONS, GPM | | | FUEL ECONOMY |
	HC	CO	NO$_x$	MPG (Liters/100km)
Full emission controls	0.35	1.41	0.35	16.2 (14.5)
No emission controls	3.13	7.00	1.46	24.3 (9.7)
Carbureted L-141 engine	4.50	73.18	3.22	15.3 (15.4)
Proposed Federal Limit (1976)	0.40	3.4	.41	
Gasoline is the fuel used.				

A centrifugal compressor is inherently a high-speed device. Even at part load, the compressor can easily be rotating 20,000 or 30,000 rpm. This high speed breaks up small droplets of gasoline and does a thorough mixing job on the already vaporized gasoline.

A correctly installed turbocharger ensures a relatively even fuel/air mixture to all cylinders. An even, or *homogeneous,* mixture promotes consistent combustion in each cylinder.

This is one of the reasons turbocharger location is so important. The compressor outlet should be located so that no stratification takes place in the intake manifold. For a further explanation see Chapter 15.

The improvement in gasoline-engine emissions is understandable, given the part-load basis on which the emissions tests are run. But the next questions are, "What about when the engine is really putting out power? Won't the emissions be a lot worse than on a high-compression naturally aspirated engine?"

Back in 1936, some gentlemen named Hershey, Eberhardt and Hottel published a graph of compression-pressure curves in the *SAE Transactions,* Figure 19-2. This graph can be used to determine compression pressure at the end of the compression stroke with a chemically correct fuel/air mixture.

Based on this graph, the compression pressures of naturally aspirated and turbocharged engines can be compared. Compression pressure has a direct effect on the combustion temperature. I've used a naturally aspirated engine with 10.5:1 compression ratio and a turbocharged engine of 7:1 compression ratio.

Table 19-2 shows that with 10 psi intake-manifold pressure (25 psia), the compression pressure of the turbocharged engine is actually 20 psi less than the naturally aspirated engine. But the turbocharged engine has approximately 50% more air and fuel and produces more than 50% more power!

Combustion pressures are no higher, so combustion temperature will be about the same. With the improved fuel distribution, exhaust emissions will usually be about the same or a little less with the turbocharged engine.

Gary Knudsen of McLaren Engines ran tests of this type on a big-block Chevy engine. He observed the compression pressure in the combustion chamber with a pressure transducer and an oscilloscope.

Knudsen was surprised to learn that the turbocharger did not appreciably increase the compression-pressure peak. The turbocharger simply fattened up the curve considerably. The fatter curve increased the bmep and therefore the horsepower.

At wide-open throttle on a system with low back pressure after the turbine, intake-manifold pressure is usually higher than exhaust-manifold pressure. This will remove the clearance gases and reduce the combustion temperature of the turbocharged engine even further.

Stratified Charge—For many years, Texaco experimented with a different type of combustion chamber using direct cylinder injection, like a diesel. The design was known as the stratified-charge, or Texaco Controlled Combustion System, TCCS.

Many SAE papers have been published on this subject, so I won't go into detail on the design. I would like to point out the addition of a turbocharger greatly increased maximum output of the engine, improved fuel economy and reduced exhaust emissions.

Table 19-3 shows where the turbocharged version of this engine equipped with full emissions controls not only met the then-proposed 1976 Federal limitations but did it at better fuel economy than the original engine with practically no emissions-control devices.

This engine might be considered halfway between a diesel and a spark-ignition engine. It has cylinder injection, which actually sprays the fuel on the spark plug. I say *fuel* because it will run on almost anything that will burn.

SUMMARY

Turbocharging definitely has a place in the engine of the future. It can improve power-to-weight ratio and help reduce exhaust emissions without hurting fuel consumption.

20 Water Injection

Most bolt-on turbocharger kits for the street produce a maximum intake-manifold pressure somewhere between 5 and 10 psig. This amount of boost normally will not require any engine modifications. Generally the engine will not detonate on premium fuel when the fuel/air mixture and ignition are correct.

Unfortunately, premium fuel at the corner gas station has gone the way of the dinosaur. Also, modifications made to allow boost pressures above 10 psig may make it impossible to prevent detonation, even with retarded spark.

In Chapter 5, I discussed enriching the fuel/air ratio as a way to prevent detonation when the engine is supercharged. Fuel enrichment will work up to a certain limit; what that limit is depends on the engine and installation.

Beyond this limit the engine will detonate regardless of how rich the mixture is. At this point, either water or water/alcohol injection is required to prevent detonation.

We all have a tendency when we discover a problem and its solution to think we have "invented the wheel." Figure 20-1 shows a device patented by Maurice Goudard in 1942. This was designed for aircraft engines, but can be adapted to others.

Even earlier, two Englishmen, Fedden and Anderson, designed a method of introducing anti-detonant to a supercharged engine. A schematic drawing of their 1936 British patent is shown in Figure 20-2. These are shown for historic reasons, but they also point out that the same thermodynamic problems exist today as did in the 1930s.

In the early 1930s, Sir Harry Ricardo ran a series of tests on detonation. Results are plotted in Figure 20-3.

In his book, *The High-Speed Internal Combustion Engine*, Ricardo notes: "In this case, running throughout at a speed of 2500 rpm and with a compression ratio of 7:1, the engine was run on an economical mixture, i.e., about 10% weak, and supercharge applied to the first incidence of

Tom Keosababian combined ecology with engineering by running on propane to break class record at Bonneville in August 1974. Car went over 170 mph that year. He used two Rajay turbochargers, two propane carburetors and water injection.

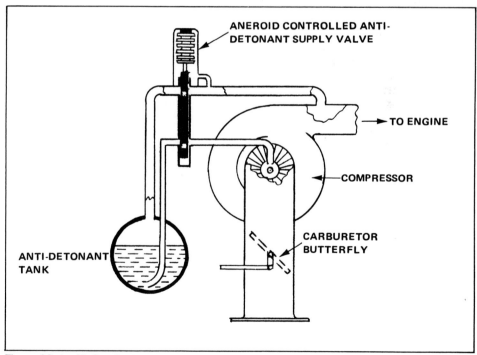

Figure 20-1—M. Goudard device Patent 2,290,610

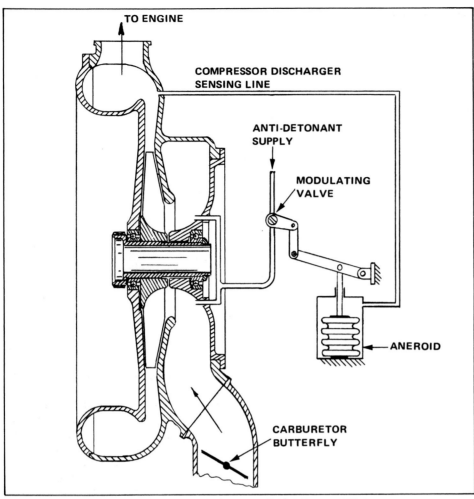

Figure 20-2—Fedden and Anderson anti-detonant, British Patent 458,611

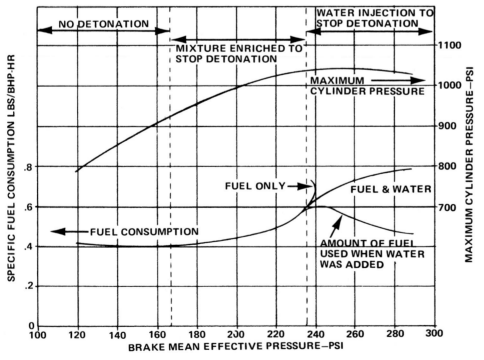

Figure 20-3—Variation in maximum bmep without detonation by enriching mixture and with water injection. Fuel was 87-octane gasoline.

detonation, which occurred when the bmep had reached 168 pounds per square inch. The mixture strength was then increased, step by step, and more supercharge applied until the same intensity of detonation was recorded; this process was continued until a point was reached at which no further enrichment was effective. In fact, after about 60% excess fuel, not only did further enrichment have no effect but there was even some indication that it increased the tendency to detonate. A finely pulverized water spray was then delivered into the induction pipe which served to suppress detonation, in part by the intercooling it provided, and in part by the influence of steam as an anti-detonant, and so allow of further supercharging. This was continued progressively, admitting just sufficient water at each stage to ward off detonation until a bmep of 290 pounds per square inch was reached, which was found to be the limit of the dynamometer. At the same time, it was noted that, with the addition of water, the influence of steam as an anti-knock allowed the fuel/air ratio being much reduced.

"From this curve Figure 20-3, it will be seen that under these operating conditions a limiting bmep that could be reached with 87-octane petrol alone at an economical mixture strength was 168 psi. By enriching the mixture to the limit of usefulness, the bmep could be stepped up to 237 psi. By the introduction of water, it could be further stepped up to 290 psi and probably more; at the same time the fuel/air ratio could be reduced once again; in fact with water injection, no appreciable advantage was found from the use of an overrich fuel/air mixture. It will be noted that the total specific consumption of liquid, i.e., fuel plus water, is not so very much greater than when running on a very rich mixture of fuel alone.

"The slope of the curve of maximum cylinder pressure is interesting in that after the injection of water, *it no longer rose but even tended very slightly to fall*, and the same applied to the gross heat flow to the cooling water which reached a maximum at a bmep of 240 pounds per square inch, and thereafter fell off, until at a bmep of 290 pounds per square inch it had fallen to the same level as that of 170 pounds without water injection."

The result of this kind of test will

134

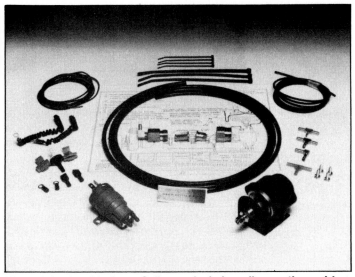

Water-injection kit from Callaway includes all mounting, wiring and plumbing hardware. Photo by C. E. Green.

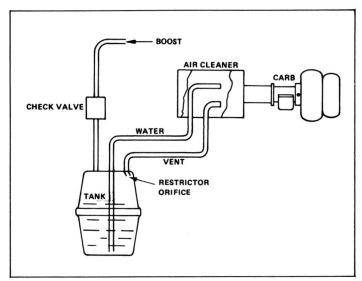

Figure 20-4—Crown water-injection system

vary with engine speed, engine size, compression ratio, combustion chamber, fuel octane etc. These variations are not important as far as the principle is concerned.

The major point is that detonation may be prevented up to a certain point by fuel enrichment. After that point, however, water or water/alcohol injection must be used.

Water & Alcohol—Water is usually combined with methanol (methylalcohol) in proportions up to 50/50. The methanol increases the volatility—and therefore the cooling effect—of the injected mixture. Methanol is also a fuel, so it gives a further horsepower increase. Finally, it eliminates the possibility of the water freezing on a cold day.

Again quoting from Ricardo's book, he notes, "Higher percentages of methanol are not desirable because methanol, itself, is prone to preignition."

Ted Trevor of Crown Manufacturing also did some testing in 1971. He showed that mixtures containing more than 50% methanol provided no additional HP gains over a 50/50 mixture. In tests done several years earlier, Dick Griffin confirmed that 50/50 is practical.

Before I discuss this further, let me say water injection is not for the average street machine. An engine with an 8:1 compression ratio and 7-psi boost should be able to use 91-octane fuel without detonation or spark retard.

In most cases, fuel enrichment alone will prevent detonation. Water or water/alcohol injection is often used to cover up bad carburetion or a poor turbocharger match.

Water-Injection Problems—Water injection can cause more problems than it solves. For example:
- If you depend on water injection and forget to fill the tank, look out.
- If you live where the temperature goes below freezing, don't forget the alcohol.
- If you have a system with an electric pump and run out of water, the motor will burn out—no more water injection, even after you refill the tank.
- If the electric pressure switch sticks closed, the pump will force all of the water into the engine. This can cause

hydraulic lock and destroy the engine the next time you try to start it.

There is the story of the fellow with a turbocharged pickup with water injection. He was cruising through the mountains at high speed. The ambient temperature was about 35F and he had no alcohol mixed with the water.

When he decided to slow down, he found the carburetor butterfly was frozen open. The temperature drop through the venturi was enough to cause icing—most embarrassing to say the least.

Water injection is great, but don't use it unless you need it. Try all the other ways of eliminating detonation first.

Types of Water Injection—At least

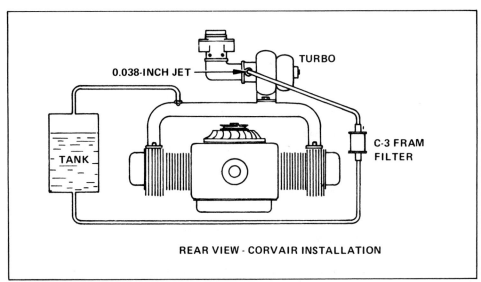

Figure 20-5—Dick Griffin water-injection system

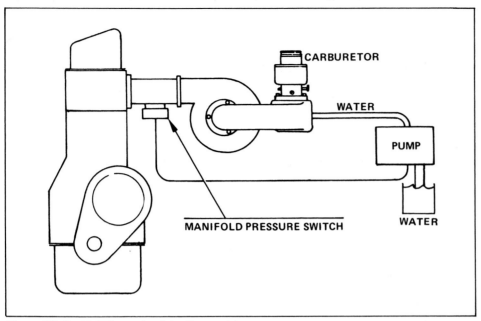

Figure 20-6—Ak Miller water-injection system

pressor inlet from a valve mounted on the top of the firewall. Carburetion was two Impco CA425 propane mixers. Keosababian added a few other goodies such as a Mallory distributor and transistor ignition.

This setup produced 450 HP at 6700 rpm and ran 176.125 mph at Bonneville in 1974. He showed it is possible to break a record without spending a fortune.

Electrical Connections—One final note: With any electrical injection system, make sure the motor is wired in series with both the pressure switch *and* the ignition switch. This will prevent any possibility of pumping water into the engine when it is not running.

DIESEL WATER INJECTION

Water/alcohol injection is not limited to spark-ignition engines. Recently, there has been much discussion about using water and/or alcohol injection on turbocharged diesels. Detonation is not a problem with a diesel, so there has to be a different reason.

Actually, there are several. Power from a highly turbocharged diesel is usually limited by exhaust temperature rather than smoke. The addition of water to the charge air reduces the peak exhaust temperature. This allows more fuel to be burned without hurting the engine.

The addition of alcohol to the water will also increase power without additional diesel fuel.

Where corn is an important crop, it is easy to justify using ethyl alcohol. It not only increases the demand for corn, but also reduces the dependence on imported oil. A similar rationale is used in Brazil, where ethyl alcohol made from sugar cane is widely used to power automobiles.

M & W Gear Company, the pioneer in turbocharging farm tractors, developed a system for diesel engines, Figure 20-7. The system injects a 50/50 mixture of water and ethanol into the inlet of the turbocharger on demand.

The system works as follows: The water/ethanol mixture is stored in a sealed tank, 4. As pressure increases in the intake manifold, 9, the tank is pressurized through a line, 1.

At a preset level—usually about 7 or 8 psig—pressure will force the fluid through a line, 5, and open the check valve, 6. The water/alcohol mixture then sprays into the inlet of the tur-

three types of water-injection systems can be considered. There are probably a lot more types, but I did not think of them as this chapter was put together.

Two types apply manifold boost pressure to the anti-detonant container. The type shown in Figure 20-4 uses pressure to push the liquid through a tube to the carburetor inlet. It uses a vent at the carburetor air inlet alongside the injector outlet jet. This vent bleeds off some pressure in the tank so injection does not start until after boost pressure has reached several psi. The exact pressure can be adjusted by the size of the restrictor in the vent line.

In the Griffin type, Figure 20-5, liquid is pushed through a fitting in the carburetor base.

A third type, by Ak Miller, uses a constant-pressure windshield-wiper pump, Figure 20-6. When boost reaches 5 psi, a pressure switch on the intake manifold turns on the pump.

Although seemingly more complicated than the other two, this system is probably the most economical. It can be made by modifying an electric windshield-washer pump. Little else is needed except a pressure switch and a tube to the carburetor.

Another system of this type could use water stored under air pressure to eliminate the pump. This would be similar to the old VW windshield washers. The system can be activated by a solenoid valve actuated at 5 psi by a pressure switch on the intake manifold.

From these variations you can develop many other combinations depending on your imagination, pocketbook, time and inclination toward experimentation.

Doug Roe used a slightly different approach to the problem on a Corvair engine. His system cooled the heads by spraying water into the cooling airstream. V-type, 3/16-inch "shooters," or spray tubes with 0.040-inch orifices, were located at the front and rear of the cooling-fan inlet. These were supplied with water when boost pressure reached 5 psi or more.

Roe's system used boost pressure to move the water. A windshield-wiper pump and manifold-pressure switch could be used instead. Also, this type of system could be augmented by using water injection into the manifold with the air/fuel mixture.

Tom Keosababian has clearly shown what can be done if water injection is used to its limit. His 1965 Corvair is a street machine with stock suspension and two stock-Corvair Rajay F turbochargers.

Water was injected into each com-

bocharger compressor. Injection continues until the intake-manifold pressure falls below the preset level.

A water/air separator, 3, at the top of the tank prevents liquid from being drawn into the engine when pressure suddenly drops in the pressure line.

Unlike some systems used on spark-ignition engines, this system is fail-safe. There is no way for the liquid to flow when the engine is not running. Also, lack of the water/alcohol mixture causes only a loss of power, not detonation.

The biggest drawback is that the water/alcohol storage tank is rather large. The actual amount injected can be as much as 30% of the amount of diesel fuel used.

The system has been used under a steady load on both farm tractors and highway trucks. Still, this and other systems of water, water/alcohol or alcohol injection for diesels are still too new to be judged. Fuel shortage threats have hastened a lot of testing along these lines.

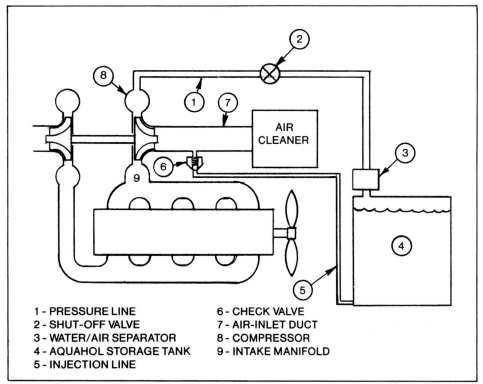

1 - PRESSURE LINE
2 - SHUT-OFF VALVE
3 - WATER/AIR SEPARATOR
4 - AQUAHOL STORAGE TANK
5 - INJECTION LINE
6 - CHECK VALVE
7 - AIR-INLET DUCT
8 - COMPRESSOR
9 - INTAKE MANIFOLD

Figure 20-7 — M & W aquahol-injection system for diesel farm tractors

21 Motorcycles

Mr. Turbo Kawasaki Funny Bike with turbocharged engine turns 1/4 mile in the 7.8-second/175-mph bracket.

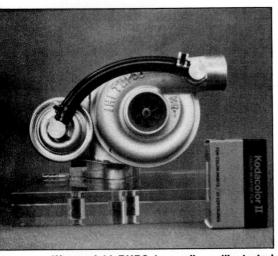

Warner-Ishi RHB3 is small, as illustrated by film pack. The problem of turbocharging small-displacement engines, such as those used in motorcycles, no longer exists. This unit will work on 0.5 to 1.3liter engines.

Turbocharging a motorcycle engine involves all the same basic principles as any other engine. But some of the problems associated with it are unique.

Until the introduction of the AiResearch T-2 and T-3, the Roto-Master RM-60, and the IHI RHB5 and RHB6 turbochargers, turbos were too large for motorcycle engines. Into the '70s, the smallest turbochargers were much too large even for the 74- and 80-CID Harley-Davidsons.

Despite this, many installations were made and gave outstanding results. Several companies are now in the business of producing bolt-on turbocharger kits.

Because of the trend to lighter, lower-displacement motorcycles, the manufacturers jumped on the turbocharger band wagon. Some of the later production turbo bikes include: Honda CX650, Kawasaki GPZ750, Suzuki XN85 and Yamaha XJ650. These bikes are in the 650 to 750cc-displacement range and are capable of 1/4-mile times of 12 seconds (plus or minus one second) at 100 to 120 mph. Because of its higher displacement—739cc—the Kawasaki GPZ750 is the fastest at 11.22 seconds at just a tick over 120 mph.

Lubrication Requirements—The amount of oil used by a turbocharger is insignificant on a passenger-car engine, even a small one. Therefore, it is rarely necessary to add any capacity to the lubrication system.

This is not so with a motorcycle engine. The capacity of a motorcycle oil sump is small—usually two or three quarts. The half gallon a minute used by a turbocharger may lower the oil pressure to the rest of the engine when it needs it most.

For this reason it is advisable to have an oil-pressure gage on the bike. If there are signs of oil starvation, install a larger-capacity oil pump.

In Chapter 9, Lubrication, I discussed the increase in oil temperature due to the additional heat from the turbocharger. This can be a serious problem on an air-cooled motorcycle. Consequently, it's a good idea to put an oil cooler on the bike.

It is possible to turbocharge a two-stroke motorcycle, but the lack of an oil pump makes it difficult. The problem is covered in Chapter 12. Two-strokes pose many more problems than four-stroke engines.

Ignition Requirements—Few motorcycles are equipped with distributors such as those on passenger cars. The vast majority have a simple mechanical or electronic device to fire the plugs. Most motorcycles fire each plug at the top of both the exhaust stroke and compression stroke. This type ignition works well on a naturally aspirated engine.

On a turbocharged engine, it's possible to have intake-manifold pressure greater than exhaust-manifold pressure. When both the exhaust valve and intake valve are open at the top of the exhaust stroke, air/fuel mixture will be blown into the combustion chamber and may be ignited by the plug.

If the mixture ignites, it will cause severe backfiring and loss of power.

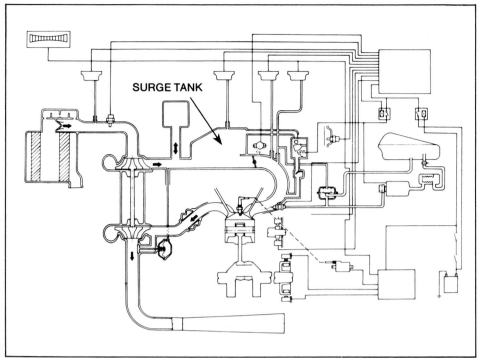

Figure 21-1—Schematic of Honda CX500 Turbo. Surge tank helps reduce pressure fluctuations between compressor and intake manifold.

The problem is rare, but if it does occur, you'll have to replace the breaker points with an automotive-type distributor or magneto. If that's too complicated, you might try a camshaft with less valve overlap.
too complicated, you might try a camshaft with less valve overlap.

Engine Types—Turbocharging motorcycles is also complicated by the number of cylinders.

A four-cylinder engine is no problem. It has an intake stroke every 180°. This gives an almost uninterrupted flow from the compressor discharge. Consequently, a four-cylinder engine doesn't require much volume between the compressor housing and the intake ports.

A two-cylinder engine, however, doesn't have an intake stroke every 180°. In fact, if it's a V-twin, it doesn't even have an intake stroke every 360°.

Regardless of the cylinder layout, the intake stroke on a twin lasts less than 180°. Air between the compressor discharge and the intake ports is stagnant half the time. During this time, it's possible for the compressor to go into surge.

When the intake valve opens at high rpm, the instantaneous engine demand can exceed the flow of the turbocharger. Intake-manifold pressure will drop off even though the average flow is far below the capacity of the compressor.

For this reason, a two-cylinder engine should have a plenum chamber on or incorporated into the intake manifold. The pipes from the compressor discharge to the cylinder head(s) should have at least twice the volume of a single cylinder. A plenum three times as large as a single cylinder is even better.

The 500cc Honda CX500 Turbo uses this kind of plenum. This 80° V-twin has a one-quart—about 950cc—chamber between the turbocharger and the intake manifold. This *surge tank* acts as a buffer to prevent pressure drop and compressor surge, Figure 21-1.

Sizing—As an example, I'll use a 74-CID Harley-Davidson V-twin. One cylinder has a displacement of 37 cubic inches, or

$$\frac{37 \text{ cu in.}}{1728 \text{ cu in.}} = 0.0214 \text{ cu ft}$$

At 6000 rpm, it will take 1/200 second to ingest 0.0214 cubic feet of air in about 180° of rotation. If the engine did this continuously, it would ingest

$$200 \times 60 \times 0.0214 \text{ cu ft/min} = 257 \text{ cfm}.$$

At 2:1 pressure ratio and a density ratio of 1.55, the compressor must flow:

$$257 \times 1.55 = 398 \text{ cfm}$$

This figure is checked on a Y-4 compressor map, the smallest of the TO4B size, Figure 21-2. This flow and pressure ratio occurs off the right side of the map, which tells you immediately that this compressor is too small.

Anti-Surge Plenum—With a 45° included angle between cylinders, cylinders fire at 360° + 45° and 360° −45°, or 405° and 315°. This leaves at least 225° of rotation with no flow into the engine. If there is no appreciable volume between the compressor and the cylinder, flow will drop to near zero at this time. The compressor will be in surge during this interval.

A cushion of air equal to three times the volume of one cylinder will lessen this problem considerably.

Assume a 2-inch-diameter compressor-discharge pipe and a volume of 3 x 37 or 111 cubic inches.

$$\text{Volume} = \frac{\pi D^2 L}{4}$$

So plenum length is calculated

$$111 = \frac{\pi 2^2 L}{4}$$

$$111 = \pi L$$

$$L = \frac{111}{\pi}$$

or approximately 35.3 in.

Increasing the diameter to 2.5 inches will reduce the length to 22.6 inches.

If the diameter is increased to 3 inches, the length goes to 15.7 inches, which does not seem unreasonable. This can probably be made part of the intake manifold.

We are all used to seeing one pipe and one carburetor for each cylinder on a naturally aspirated bike. It may be hard for the uninitiated to believe an exhaust system with all the cylinders connected into a single pipe and an intake system with a single carburetor can give much better performance just because of a little gadget called a *turbocharger*. I can assure you this is the case.

One of the most annoying things about some modified bikes is the noise—particularly if you travel alongside one for any length of time. The combination of joining the exhaust

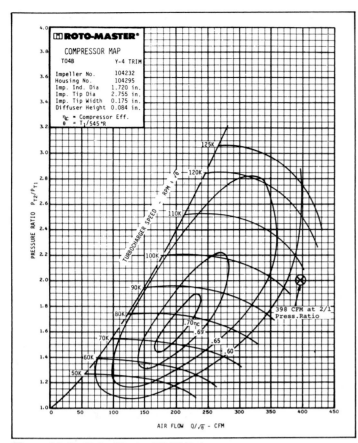

Figure 21-2—Roto-Master T04B Y-4 compressor map

Figure 21-3—Luftmeister turbocharger installation looks at home on BMW engine—RHB6 Warner-Ishi is used. Note stock-appearing exhausts.

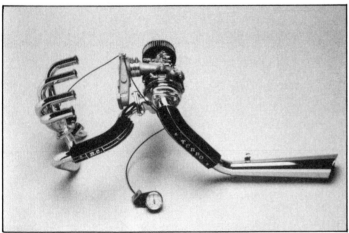

RC Engineering kit for Kawasaki Z-1/KZ1000 uses Rajay turbocharger and waste gate with a specially built Keihin carburetor. Company also has kits for Suzuki GS750/850/1000 and Honda 750 fours.

Matt Capri and Lou Nauert of Luftmeister posing with their 1977 R100 turbocharged BMW. Bike ran a record 10.50 seconds at 139 mph in the 1/4 mile. It is claimed the bike is capable of 175-mph top speed with proper gearing and a *lot of road.*

pipes together and running them through the turbocharger does an excellent job of quieting the exhaust.

The turbocharged motorcycle with no muffler is usually quieter than a naturally aspirated one with a muffler on each pipe. This might not appeal to some, but it is pleasing to anyone driving behind or alongside.

Figure 21-3 shows the BMW designers must have had turbocharging in mind because it fits in so neatly. Turbos are not obvious on most motorcycle installations. The only thing noticeable as it drives by is the single exhaust pipe.

Cline Turbo Technology's kit, manufactured by Blake Enterprises, has several kits including one for the Suzuki GS 1100. Their kit for the stock Z-1 Kawasaki starts to get boost at 5000 rpm and reaches 18 psi at 8500 rpm. It easily pulls 10,000 rpm in fifth gear. Depending on fairing, gearing and tires, it will run 150 to 190 mph in street trim.

This kit goes all the way with water/alcohol injection, pressure-retarded spark and a separate oil pump and filter. Any one for a Sunday drive in the country?

Appendix

Glossary

Adiabatic—Ideal reversible compression of a gas. Implies that there is no heat loss in the process.

Blade—Part of compressor impeller or turbine wheel acting upon or acted upon by air or exhaust gas.

Blower—Term often applied to all types of superchargers, but usually to low-pressure-ratio units.

Blowoff valve—A spring-loaded valve in either intake or exhaust manifold to prevent overboosting.

Boost—Difference between barometric pressure and intake-manifold pressure on a supercharged engine.

Charge—Air/fuel mixture to be burned in combustion chamber.

Clearance volume—The volume of a cylinder between the top of the piston and the cylinder head.

Combustion chamber—Volume between cylinder head and top of piston when piston is at top of stroke.

Compression ratio—Ratio of clearance volumes of a cylinder at top and bottom dead center.

Compression—That portion of the turbocharger that increases the pressure of the air or air/fuel mixture.

Compressor-flow range—That area of the compressor map between the surge line and 60% efficiency (for centrifugal compressors only).

Compressor housing—The housing that encloses the compressor. Sometimes referred to as a *scroll.*

Compressor impeller—The rotating portion of the compressor.

Compressor-pressure ratio—Compressor-outlet absolute pressure divided by compressor-inlet absolute pressure.

Critical gas flow—Maximum flow through an orifice or restriction for a given pressure upstream of the orifice or restriction.

Density—Weight of air or charge per cubic foot of volume.

Diffuser—The stationary portion of the compressor that increases the static pressure of the air or air/fuel mixture.

Efficiency—Actual performance of a piece of equipment compared to the idea.

Enthalpy—Internal energy of a working fluid. Usually stated in BTU/lb.

Exducer—The gas-exit portion of a radial turbine wheel.

Exhaust gas recirculation (EGR)—System used to duct some engine exhaust back into intake manifold. Reduces oxides of nitrogen in exhaust.

Fin—Thin projection on inside of hollow casting caused by crack in core.

Inducer—The gas-entry portion of a centrifugal-compressor rotor.

Intercooler (aftercooler or heat exchanger)—A heat exhanger that reduces the temperature of the compressed charge before it enters the combustion chamber.

Micron—Particle-size measurement used to indicate quality of air or oil filter. 1 micron = 0.00004 inch.

Naturally aspirated (NA)—An engine without a supercharger.

Normalize—Supercharge an engine running at high altitude, but only to regain power lost because of lower air density.

Nozzle—Stationary portion of the turbine that increases velocity of exhaust gases and directs them to the turbine wheel. Most small turbochargers accomplish this with a scroll-shaped turbine housing.

Positive crankcase ventilation (PCV)—System that ducts engine crankcase fumes back to the intake manifold to reduce air pollution.

Pressure, absolute—Pressure measured above a complete vacuum.

Pressure, boost—Intake-manifold gage pressure. 1-psi boost = 2-in.Hg boost (approximately).

Pressure, gage—Pressure measured between two places, usually between ambient and manifold.

Pressure, static—Pressure measured in a housing or duct through a hole in a wall that is parallel to the direction of flow.

Pressure, total—Pressure measured in a housing or duct with a probe that senses the velocity pressure as well as the static pressure.

Residual gases—Exhaust gases left in the clearance volume at the end of the exhaust stroke.

Rotor—Rotating portion of a turbocharger, including the impeller, shaft and turbine wheel.

Scavenging—Removing combustion products from combustion chamber.

Stoiciometric mixture—Correct chemical mixture of fuel and air for complete combustion of both.

Stroke—Distance traveled by piston between top and bottom dead center.

Supercharge—Increases density of charge by compressing it before it enters the combustion chamber.

Surge line—A line on a centrifugal-compressor map representing minimum flow at each pressure ratio.

Throat—Entry portion of turbine housing that defines nozzle area.

Torus—Doughnut shaped compressor housing sometimes called *collector-type housing.*

Turbine—Portion of the turbocharger that converts energy of the engine exhaust gases to shaft power.

Turbine wheel—The rotating portion of the turbine.

Turbocharger—An engine supercharger driven by an exhaust-gas turbine.

Valve overlap—The number of crankshaft degrees expressing the time when both the intake and exhaust valves are open.

Vane—Stationary guide of diffuser or nozzle.

Volute—A scroll or snail shaped housing.

Vortex—Free-flowing inward spiral, like that at the drain of a bathtub.

Y Tables

VALUES OF "Y" FOR NORMAL AIR AND PERFECT DITOMIC GASES

$$Y = r^{.283} - 1 \qquad r = P_{T2}/P_{T1} \qquad K = 1.395$$

r		0	1	2	3	4	5	6	7	8	9
1.00	0.00	000	028	057	085	113	141	169	198	226	254
1.01		282	310	338	366	394	422	450	478	506	534
1.02		562	590	618	646	673	701	729	757	785	812
1.03		840	868	895	923	951	978	006	034	061	089
1.04	0.01	116	144	171	199	226	253	281	308	336	363
1.05		390	418	445	472	500	527	554	581	608	636
1.06		663	690	717	744	771	798	825	852	879	906
1.07		933	960	987	014	041	068	095	122	148	175
1.08	0.02	202	229	255	282	309	336	362	389	416	442
1.09		469	495	522	549	575	602	628	655	681	708
1.10		734	760	787	813	840	866	892	919	945	971
1.11		997	024	050	076	102	129	155	181	207	233
1.12	0.03	258	285	311	337	363	389	415	441	467	493
1.13		519	545	571	597	623	649	675	700	726	752
1.14		778	804	829	855	881	906	932	958	983	009
1.15	0.04	035	060	086	111	137	162	188	213	239	264
1.16		290	315	341	366	391	417	442	467	493	518
1.17		543	569	594	619	644	670	695	720	745	770
1.18		796	821	846	871	896	921	946	971	996	021
1.19	0.05	046	071	096	121	146	171	196	221	245	270
1.20		295	320	345	370	394	419	444	469	493	518
1.21		543	567	592	617	641	666	691	715	740	764
1.22		789	813	838	862	887	911	936	960	985	009
1.23	0.06	034	058	082	107	131	155	180	204	228	253
1.24		277	301	325	350	374	398	422	446	470	495
1.25		519	543	567	591	615	639	663	687	711	735
1.26		759	783	807	831	855	879	903	927	951	974
1.27		998	022	046	070	094	117	141	165	189	212
1.28	0.07	236	260	283	307	331	354	378	402	425	449
1.29		472	496	520	543	567	590	614	636	661	684
1.30		708	731	754	778	801	825	848	871	895	918
1.31		941	965	988	011	035	058	081	104	128	151
1.32	0.08	174	197	220	243	267	290	313	336	359	382
1.33		405	428	451	474	497	520	543	566	589	612
1.34		635	658	681	704	727	750	773	795	818	841
1.35		864	887	910	932	955	978	001	023	046	069
1.36	0.09	092	114	137	160	182	205	228	250	273	295
1.37		318	341	363	386	408	431	453	476	498	521
1.38		543	566	588	611	633	655	678	700	723	745
1.39		767	790	812	834	857	879	901	923	946	968
1.40		990	012	035	057	079	101	123	145	168	190
1.41	0.10	212	234	256	278	300	322	344	366	389	411
1.42		433	455	477	499	521	542	564	586	608	630
1.43		652	674	696	718	740	761	783	805	827	849
1.44		871	892	914	936	958	979	001	023	045	066
1.45	0.11	088	110	131	153	175	198	218	239	261	283
1.46		304	326	347	369	390	412	433	455	476	498
1.47		520	541	562	584	605	627	648	669	691	712
1.48		734	755	776	798	819	840	862	883	904	925
1.49		947	968	989	010	032	053	074	095	116	138
1.50	0.12	159	180	201	222	243	264	286	307	328	349
1.51		370	391	412	433	454	475	496	517	538	559
1.52		580	601	622	643	664	685	706	726	747	768
1.53		789	810	831	852	872	893	914	935	956	977
1.54		997	018	039	060	080	101	122	142	163	184
1.55	0.13	205	225	246	266	287	308	328	349	370	390
1.56		411	431	452	472	493	513	534	554	575	595
1.57		616	636	657	677	698	718	739	759	780	800
1.58		820	841	861	881	902	922	942	963	983	003
1.59	0.14	024	044	064	085	105	125	145	165	186	206
1.60		226	246	267	287	307	327	347	367	387	408
1.61		428	448	468	488	508	528	548	568	588	608
1.62		628	648	668	688	708	728	748	768	788	808
1.63		828	848	868	888	908	928	948	968	988	007
1.64	0.15	027	047	067	087	107	126	146	166	186	206
1.65		225	245	265	284	304	324	344	363	383	403
1.66		423	442	462	481	501	521	540	560	580	599
1.67		619	638	658	678	697	717	736	756	775	795
1.68		814	834	853	873	892	912	931	951	970	990
1.69	0.16	009	028	048	067	087	106	125	145	164	184
1.70		203	222	242	261	280	299	319	338	357	377
1.71		396	415	434	454	473	492	511	531	550	569
1.72		588	607	626	646	665	684	703	722	741	760
1.73		780	799	818	837	856	875	894	913	932	951
1.74		970	989	008	027	046	065	084	103	122	141

r		0	1	2	3	4	5	6	7	8	9
1.75	0.17	160	179	198	217	236	255	274	292	311	330
1.76		349	368	387	406	425	443	462	481	500	519
1.77		538	556	575	594	613	631	650	669	688	706
1.78		725	744	762	781	800	818	837	856	874	895
1.79		912	930	949	968	988	005	023	042	061	079
1.80	0.18	098	116	135	153	172	191	209	228	246	265
1.81		283	302	320	339	357	376	394	412	431	449
1.82		468	486	505	523	541	560	578	596	615	633
1.83		652	670	688	707	725	743	762	780	798	816
1.84		835	853	871	890	908	926	944	962	981	999
1.85	0.19	017	035	054	072	090	108	126	144	163	181
1.86		199	217	235	253	271	289	308	326	344	362
1.87		380	398	416	434	452	470	488	506	524	542
1.88		560	578	596	614	632	650	668	686	704	722
1.89		740	758	776	794	811	829	847	865	883	901
1.90		919	937	954	972	990	008	026	044	061	079
1.91	0.20	097	115	133	150	168	186	204	221	239	257
1.92		275	292	310	328	345	363	381	399	416	434
1.93		452	469	487	504	522	540	557	575	593	610
1.94		628	645	663	681	698	716	733	751	768	786
1.95		804	821	839	856	874	891	909	926	944	961
1.96		979	996	013	031	048	066	083	101	118	135
1.97	0.21	153	170	188	205	222	240	257	275	292	309
1.98		327	344	361	379	396	413	431	448	465	482
1.99		500	517	534	552	569	586	603	620	638	655
2.00		672	689	707	724	741	758	775	792	810	827
2.01		844	861	878	895	913	930	947	964	981	998
2.02	0.22	015	032	049	066	084	101	118	135	152	169
2.03		186	203	220	237	254	271	288	305	322	339
2.04		356	373	390	407	424	441	458	474	491	508
2.05		525	542	559	576	593	610	627	644	660	677
2.06		694	711	728	745	761	778	795	812	829	846
2.07		863	879	896	913	930	946	963	980	997	013
2.08	0.23	030	047	064	080	097	114	130	147	164	181
2.09		197	214	231	247	264	281	297	314	331	347
2.10		364	380	397	414	430	447	463	480	497	513
2.11		530	546	563	579	596	613	629	646	662	679
2.12		695	712	728	745	761	778	794	811	827	844
2.13		860	877	893	909	926	942	959	975	992	008
2.14	0.24	024	041	057	074	090	106	123	139	155	172
2.15		188	204	221	237	253	270	286	302	319	335
2.16		351	368	384	400	416	433	449	465	481	498
2.17		514	530	546	563	579	595	611	627	644	660
2.18		676	692	708	724	741	757	773	789	805	821
2.19		838	854	870	886	902	918	934	950	966	983
2.20		999	015	031	047	063	079	095	111	127	143
2.21	0.25	159	175	191	207	223	239	255	271	287	303
2.22		319	335	351	367	383	399	415	431	447	463
2.23		479	495	511	526	542	558	574	590	606	622
2.24		638	654	669	685	701	717	733	749	765	780
2.25		796	812	828	844	859	875	891	907	923	938
2.26		954	970	986	001	017	033	049	064	080	096
2.27	0.26	112	127	143	159	175	190	206	222	237	253
2.28		269	284	300	316	331	347	363	378	394	409
2.29		425	441	456	472	488	503	519	534	550	566
2.30		581	597	612	628	643	659	675	690	706	721
2.31		737	752	768	783	799	814	830	845	861	876
2.32		892	907	923	938	954	969	984	000	015	031
2.33	0.27	046	062	077	092	108	123	139	154	169	185
2.34		200	216	231	246	262	277	292	308	323	338
2.35		354	369	384	400	415	430	446	461	476	492
2.36		507	522	538	553	568	583	599	614	629	644
2.37		660	675	690	705	721	736	751	766	781	797
2.38		812	827	842	857	873	888	903	918	933	948
2.39		964	979	994	009	024	039	054	070	085	100
2.40	0.28	115	130	145	160	175	190	205	220	236	251
2.41		266	281	296	311	326	341	356	371	386	401
2.42		416	431	446	461	476	491	506	521	536	551
2.43		566	581	596	611	626	641	656	671	686	701
2.44		716	730	745	760	775	790	805	820	835	850
2.45		865	879	894	909	924	939	954	969	984	998
2.46	0.29	013	028	043	058	073	087	102	117	132	147
2.47		162	176	191	206	221	235	250	265	280	295
2.48		309	324	339	353	368	383	398	412	427	442
2.49		457	471	486	501	515	530	545	559	574	589
2.50		604	618	633	647	662	677	691	706	721	735
2.51		750	765	780	794	808	823	838	852	867	881
2.52		896	911	925	940	954	969	984	998	013	027
2.53	0.30	042	056	071	085	100	114	129	144	158	173
2.54		187	202	216	231	245	260	274	289	303	318
2.55		332	346	361	375	390	404	419	433	448	462
2.56		476	491	505	520	534	548	563	577	592	606
2.57		620	635	649	663	678	692	707	721	735	750
2.58		764	778	793	807	821	836	850	864	879	893

r		0	1	2	3	4	5	6	7	8	9
2.59		907	921	936	950	964	979	993	007	021	036
2.60	0.31	050	064	079	093	107	121	136	150	164	178
2.61		193	207	221	235	249	264	278	292	306	320
2.62		335	349	363	377	391	405	420	434	448	462
2.63		476	490	505	519	533	547	561	575	589	603
2.64		618	632	646	660	674	688	702	716	730	744
2.65		759	773	787	801	815	829	843	857	871	885
2.66		899	913	927	941	955	969	983	997	011	025
2.67	0.32	039	053	067	081	095	109	123	137	151	165
2.68		179	193	207	221	235	249	262	276	290	304
2.69		318	332	346	360	374	388	402	416	429	443
2.70		457	471	485	499	513	527	540	554	568	582
2.71		596	610	624	637	651	665	679	693	707	720
2.72		734	748	762	776	789	803	817	831	845	858
2.73		827	886	900	913	927	941	955	968	982	996
2.74	0.33	010	023	037	051	065	078	092	106	119	133
2.75		147	161	174	188	202	215	229	243	256	270
2.76		284	297	311	325	338	352	366	379	393	407
2.77		420	434	448	461	475	488	502	516	529	543
2.78		556	570	584	597	611	624	638	651	665	679
2.79		692	706	719	733	746	760	773	787	801	814
2.80		828	841	855	868	882	895	909	922	936	949
2.81		963	976	990	003	017	030	044	057	070	084
2.82	0.34	097	111	124	138	151	165	178	191	205	218
2.83		232	245	259	272	285	299	312	326	339	352
2.84		366	379	393	406	419	433	446	459	473	486
2.85		500	513	526	540	553	566	580	593	606	620
2.86		633	646	660	673	686	700	713	726	739	753
2.87		766	779	793	806	819	832	846	859	872	886
2.88		899	912	925	939	952	965	978	991	005	018
2.89	0.35	031	044	058	071	084	097	110	124	137	150
2.90		163	176	190	203	216	229	242	255	269	282
2.91		295	308	321	334	347	361	374	387	400	413
2.92		426	439	452	466	479	492	505	518	531	544
2.93		557	570	584	597	610	623	636	649	662	675
2.94		688	701	714	727	740	753	767	780	793	806
2.95		819	832	845	858	871	884	897	910	923	936
2.96		949	962	975	988	001	014	027	040	053	066
2.97	0.36	079	092	105	118	131	144	157	169	182	195
2.98		208	221	234	247	260	273	286	299	312	324
2.99		337	350	363	376	389	402	415	428	440	453
3.00	0.3	647	659	672	685	698	711	723	736	749	761
3.1		774	786	799	811	824	836	849	861	874	886
3.2		898	911	923	935	947	959	971	984	996	008
3.3	0.4	020	032	044	056	068	080	091	103	115	127
3.4		139	150	162	174	186	197	209	220	232	244
3.5		255	267	278	290	301	313	324	335	347	358
3.6		369	380	392	403	414	425	437	448	459	470
3.7		481	492	503	514	525	536	547	558	569	580
3.8		591	602	612	623	634	645	656	666	677	688
3.9		698	709	720	730	741	752	762	773	783	794
4.0		804	815	825	835	846	856	867	877	887	898
4.1		908	918	928	939	949	959	970	980	990	000
4.2	0.5	010	020	030	040	050	060	070	080	090	100
4.3		110	120	130	140	150	160	170	179	189	199
4.4		209	219	228	238	248	258	267	277	287	296
4.5		306	318	325	335	344	354	363	373	382	392
4.6		401	411	420	430	439	449	458	467	477	486
4.7		495	505	514	523	533	542	551	560	570	579
4.8		588	597	606	616	625	634	643	652	661	670
4.9		679	688	697	706	715	724	733	742	751	760
5.0		769	778	787	796	805	814	822	831	840	849
5.1		858	867	875	884	893	902	910	919	928	936
5.2		945	954	962	971	980	988	997	006	014	023
5.3	0.6	031	040	048	057	065	074	082	091	099	108
5.4		116	125	133	142	150	159	167	175	184	192
5.5		200	209	217	225	234	242	250	258	267	275
5.6		283	291	300	308	316	324	332	340·	349	357
5.7		365	373	381	389	397	405	413	421	430	438
5.8		446	454	462	470	478	486	494	502	509	517
5.9		525	533	541	549	557	565	573	581	588	596
6.0		604	612	620	628	635	643	651	659	666	674
6.1		682	690	697	705	713	721	729	736	744	752
6.2		759	767	774	782	789	797	805	812	820	827
6.3		835	843	850	858	865	873	880	888	895	903
6.4		910	918	925	933	940	948	955	963	970	978
6.5		985	992	000	007	014	021	028	036	043	050
6.6	0.7	058	065	073	080	087	095	102	110	117	124
6.7		131	138	145	153	160	167	174	181	189	196
6.8		203	210	217	224	232	239	246	253	260	267
6.9		274	281	288	295	302	309	316	323	330	338

r		0	1	2	3	4	5	6	7	8	9
7.0		345	352	359	366	373	380	386	393	400	407
7.1		414	421	428	435	442	449	456	463	470	477
7.2		483	490	497	504	511	518	524	531	538	545
7.3		552	559	565	572	579	586	592	599	606	613
7.4		620	626	633	640	646	653	660	666	673	680
7.5		687	693	700	706	713	720	726	733	740	746
7.6		753	760	766	773	779	786	792	799	806	812
7.7		819	825	832	838	845	851	858	864	871	877
7.8		884	890	897	903	910	916	925	929	936	942
7.9		949	955	961	968	974	981	987	993	000	006
8.0	0.8	013	019	025	032	038	044	051	057	063	070
8.1		076	082	089	095	101	108	114	120	126	133
8.2		139	145	151	158	164	170	176	183	189	195
8.3		201	207	214	220	226	232	238	245	251	257
8.4		263	269	275	281	288	295	300	306	312	318
8.5		324	330	336	343	349	355	361	367	373	379
8.6		385	391	397	403	409	415	421	427	433	439
8.7		445	451	457	463	469	475	481	487	493	499
8.8		505	511	517	523	529	535	541	547	552	558
8.9		564	570	576	582	588	594	600	605	611	617
9.0		623	629	635	641	646	652	658	664	670	676
9.1		681	687	693	699	705	710	716	722	728	734
9.2		739	747	751	757	762	768	774	779	785	791
9.3		797	802	808	814	819	825	831	837	842	848
9.4		854	859	865	871	876	882	888	893	899	905
9.5		910	916	921	927	933	938	944	949	955	961
9.6		966	972	977	983	989	994	000	005	011	016
9.7	0.9	022	028	033	039	044	050	055	061	066	072
9.8		077	083	088	094	099	105	110	116	121	171
9.9		132	138	143	149	154	159	165	170	176	181
10.0		187	192	198	203	208	214	219	225	230	235

From "Engineering Computations for Air and Gases" by Moss and Smith, Transactions A.S.M.E., Vol. 52, 1930, Paper APM-52-8

EXAMPLE:

Assume: $P_{T2}/P_{T1} = r = 1.82$

Air Flow = 290 CFM

T_1 (inlet temperature = 92F = (92 + 460)R = 552R

η_c (from Figure 5) = 68%

Y (from table) = 0.18468

T_i (ideal temperature rise) = $T_1 \times Y$ = 552 x 0.18468 = 102

ΔT_A (actual temperature rise) = $\dfrac{\Delta T_i}{\eta_c}$ = $\dfrac{102}{0.68}$ = 150F

T_2 (compressor outlet temperature) = $T_1 + \Delta T_A$ = 92 + 150 = 242F

Symbols

Symbol	Name	Used For	Units
A		Acceleration	in./sec.^{2} or ft./sec.^{2}
A		Area	Sq. in.
ABDC	After BDC	Valve timing vs. piston position	Degrees ($^\circ$)
ABS	Absolute	Pressure or temperature above absolute zero	
A/R	Area Ratio	Turbine housing size	In.
ATDC	After TDC	Valve timing vs. piston position	Degrees ($^\circ$)
BDC	Bottom Dead Center	Piston position/valve timing	
BBDC	Before BDC	Valve timing vs. piston position	Degrees ($^\circ$)
BSFC	Brake Specific Fuel Consumption		Lb/BHP-Hr.
BTDC	Before TDC	Valve timing vs. piston position	Degrees ($^\circ$)
BTU	British Thermal Unit	Unit of energy	778 Ft. Lbs.
cc	Cubic Centimeters	Engine size	Cu. cm.
CFM		Volume flow	Cubic feet per minute
CID	Cubic Inch Displacement	Engine size	Cu. in.
$\triangle$	Delta	Differences	none
e.t.	Elapsed time	Timing	Sec.
η	Eta	Efficiency	none
F	Fahrenheit	Temperature	Degrees ($^\circ$)
F	Force	Calculating horsepower	Lb.
FPS		Speed	Feet per second
G		Acceleration of gravity	384 in./sec.^{2} or 32.2 ft./sec.^{2}
HP	Horsepower	Engine output	$\dfrac{550 \text{ ft. lb.}}{\text{sec.}}$
In. Hg	Inches mercury	Pressure	Inches
I		Moment of inertia	Lb. in. Sec.^{2}
K	Radius of gyration	Moment of inertia	In.
L	Liter	Displacement	Liters
M	Slug	Mass	$\dfrac{\text{Lb. sec.}^2}{\text{ft.}}$
M.E.P.	Mean Effective Pressure	Calculating torque	psi
MPH		Speed	Miles per hour
N	Rotational speed		RPM
N.A.	Naturally aspirated	Engine designation	
P		Pressure	Lb./in^2 or In. Hg (mercury)
PSIA		Absolute pressure	Lb./in.^{2} Absolute
PSIG		Gage pressure	Lb./in.^{2} gage
Q	Volume rate of flow		Ft^3/Min.
r		Pressure ratio	none
R		Radius	In.
R	Rankine	Temperature (absolute)	Degrees ($^\circ$)
T		Temperature	$^\circ$F or $^\circ$R
θ	THETA	Temperature ratio	$\dfrac{T^\circ R}{520}$
TDC	Top Dead Center	Piston position/valve timing	
T		Torque	Lb. ft.
TC	Turbocharged	Engine designation	
v		Volume	Cu. ft. or cu. in.
V		Velocity	Ft/Sec.
W		Weight	Lbs.
Y		Calculating compressor temperature rise	

Altitude Chart

Altitude Ft.	Air Press. in.Hg	$^\circ$F	Standard Day $^\circ$R	$\sqrt{\theta}$
Sea Level	29.92	59.00	518.69	1.00
1000	28.86	55.43	515.12	.997
2000	27.82	51.87	511.56	.993
3000	26.81	48.30	507.99	.989
4000	25.84	44.74	504.42	.986
5000	24.90	41.17	500.86	.982
6000	23.98	37.61	497.30	.979
7000	23.09	34.05	493.73	.975
8000	22.23	30.48	490.17	.972
9000	21.39	26.92	486.61	.969
10000	20.58	23.36	483.04	.965
11000	19.80	19.79	479.48	.962
12000	19.03	16.23	475.92	.958
13000	18.30	12.67	472.36	.954
14000	17.58	9.11	468.80	.951
15000	16.89	5.55	465.23	.947
16000	16.22	1.99	461.67	.943
17000	15.58	− 1.58	458.11	.940
18000	14.95	− 5.14	454.55	.936
19000	14.35	− 8.69	450.99	.933
20000	13.76	−12.25	447.43	.929
21000	13.20	−15.81	443.87	.925
22000	12.65	−19.37	440.32	.921
23000	12.12	−22.93	436.76	.918
24000	11.61	−26.49	433.20	.914
25000	11.12	−30.05	429.64	.910
26000	10.64	−33.60	426.08	.906
27000	10.18	−37.16	422.53	.903
28000	9.741	−40.72	418.97	.899
29000	9.314	−44.28	415.41	.895
30000	8.903	−47.83	411.86	.891
31000	8.506	−51.38	408.30	.887
32000	8.124	−54.94	404.75	.883
33000	7.756	−58.50	401.19	.880
34000	7.401	−62.05	397.64	.876
35000	7.060	−65.61	394.08	.872
36000	6.732	−69.16	390.53	.868
37000	6.417	−69.70	389.98	.867
38000	6.117	−69.70	389.98	.867
39000	5.831	−69.70	389.98	.867
40000	5.558	−69.70	389.98	.867

Equivalents

1 N (Newton)	= 0.2248 lb force
1 Nm (Newton meter)	= 0.7376 lb ft.
1 kPa (kilo Pascal)	= 0.1450 lb/in.^{2}
1 kPa	= 0.2961 in. Hg.
1 kW ((kiloWatt)	= 1.341 HP
1 km (kilometer)	= 0.621 mile
1 cm (centimeter)	= 0.394 inch
1 cm^2	= 0.155 inch^2
1 cm^3	= 0.161 inch^3
1 l (liter)	= 61.02 inch^3
Temp. C (Celsius)	= 5/9 (TF−32)
1 km/h (kilometer/hour)	= 0.621 mph (miles per hour)
1 liter/100 km	= 235.22 miles/gal.
1 BTU	= 1.055 kJ

Turbocharger—Failure Analysis

by Robert Elmendorf
Schwitzer Service Bulletin

Despite the outward appearance of utter simplicity, the modern diesel turbocharger is, in fact, the end product of a highly specialized technology, and though not readily apparent on first observation, the methods used in turbocharger manufacture have been brought to a high level of sophistication, particularly during the past two decades.

The greatest demands on this technology are made in the areas of metallurgy, dimensional tolerance control, and dynamic unbalance correction. The reasons for extreme emphasis in these areas are easy to appreciate, once the typical stresses acting on a turbocharger are known. In normal operation, then, gas introduction temperatures can easily exceed 1000F, rotative speeds can run above 70,000 rpm, and the internal power of the turbocharger can approach values equivalent to the flywheel horsepower of the engine.

It is known even to most neophytes that turbocharger failures are inherently expensive, and have a devastating effect on engine performance. It is a widespread misconception even among the seasoned "experts," however, that all turbocharger failures occur instantaneously of primarily internal causes, and are therefore unpreventable; on the contrary, the vast majority of failures are the direct result of faulty engine maintenance or operation, and many are progressive in nature.

It is the purpose of this bulletin to acquaint the owner or operator of a "turbo"-equipped engine with the several categories of common failure, to help him recognize the symptoms associated with each, and hopefully, to enable him to avoid the expense and aggravation of repetitive failures. However, to develop a complete and well-rounded turbocharger service program, it is suggested that the information contained herein be supplemented with the wealth of routine and preventive maintenance information available for both the complete engine and the turbocharger assembly alone.

FAILURES RELATED TO FAULTY LUBRICATION

Oil plays a vital role in the life of a turbocharger, because it serves the triple function of lubricating, cooling and cleansing many of the most critical and highly stressed parts in the assembly.

Even momentary interruptions in the supply of high quality lubrication can produce disastrous results, but particularly under conditions of high speed or heavy load.

There are several facets to proper turbocharger lubrication, and as a consequence, a deficiency of any one aspect usually produces a specific symptom.

Abrasive Contamination—The presence of sufficient abrasive material in the lubricating oil will result in wear of the various bearing surfaces, and is usually most prevalent on the thrust bearing and outside diameters of the shaft bearings. Occasionally, the abrasive particles are so small that they escape the centrifuge effect of the rapidly spinning bearings; in such a case, considerable scouring of the journal sections of the rotor shaft might be noted.

The depth of scratching and amount of wear can vary widely, and depend primarily on the degree and nature of the contaminant, the operating time accumulated with the contaminant present, and the severity of engine operation.

Insufficient Pressure or Flow—It is essential that a sufficient quantity of oil flows through the turbocharger at all times to ensure suspension and stabilization of the full.floating/rotating bearing system, and to continually wash heat from the unit, thereby keeping internal temperatures within workable limits.

The most frequently seen damage that can be related to insufficient oil flow is the result of what might be termed the "oil lag syndrome." This involves a faulty operating habit, in which a cold turbocharged engine is started and immediately taken to a condition of high speed or heavy load. In such a case, the rotative speed of the turbocharger approaches the peak allowable value before an effective oil cushion is established to sustain the bearings. The result is rotor gyration (often called "shaft motion" or "whirl") with attendant distress of the bearings.

Marginal oil flow—that in which sufficient oil is circulating to prevent a sudden failure—can also produce detrimental effects, the most notable of which is the gradual accumulation of varnish on internal surfaces, which makes unit disassembly quite difficult.

Improper Oil Type or Lack of Change at Recommended Levels—The principal condition noted when either of these conditions exist is generalized sludge or varnish formation on the internal surfaces of the turbocharger; these are usually found to be heaviest at the turbine end of the unit, because the higher prevailing temperature in that area results in accelerated loss of the volatile oil components.

In certain applications, this varnish formation causes eventual seal ring fouling and wear, and can only be corrected by complete unit teardown and replacement of the turbine wheel-and-shaft assembly; this is a very expensive proposition, at best.

FAILURES RELATED TO FOREIGN MATERIAL INTAKE

The vulnerability of a turbocharger will become instantly apparent the first time a particle of significant size is inducted into either the compressor or turbine section with the unit at speed.

The sources and types of air- and exhaust-stream contamination are many and varied, but can range from atmospheric sand and dust (through the compressor) to engine valve fragments (through the turbine).

The point of foreign material entry usually becomes apparant as soon as the core has been separated from the compressor cover and turbine housing; though the damaging particle is seldom present or intact, some clue as to its size and type can usually be gained by close inspection of the involved wheel.

The secondary effects of high speed particle impact on either wheel are usually visible throughout the unit, but tend to focus on the bearings, which suffer from both the initial gyration and the running unbalance condition which follows.

FAILURES RELATED TO HIGH EXHAUST TEMPERATURES OR UNIT OVERSPEED

The procedure employed in the matching of a turbocharger to a particular engine is part of the specialized technology mentioned in the introduction to this bulletin, and always includes a live proofing session in the lab, using an actual engine under closely controllable conditions.

The reasons for this final precautionary measure are simple: As a free-wheeling device with intense and tremendous internal power, a mismatched turbocharger could easily "run away," damaging the engine and creating a threat to the well-being of the bystanders and attendants.

As seen in the field, mismatched engine/turbocharger combinations are most often the result of an attempt to boost power output by means of various "adjustments," such as fuel rate tampering, indiscriminate compressor or turbine nozzle changes, or even the application of complete turbochargers to engines arbitrarily. Any of these attempts can cause overspeed damage to the bearing system of a turbocharger, distortion, warpage or erosion of the turbine wheel casting, and heat damage to such other engine components as piston crowns, valves and exhaust manifolds.

Acknowledgements

If this book were about reciprocating engines or gas turbines, a good portion of the information could have been gleaned from existing literature because of the many books and papers on these subjects. Although a few technical papers have been published, modern books on turbocharging are almost nonexistent. Turbocharger development has progressed over the last 35 years, mainly because of better materials. For this reason and because thermodynamic principles do not change, the work done by Buchi in the early 1900s and Ricardo in the 1920-1930s is still valid.

AiResearch, Rajay, Roto-Master, Schwitzer and Warner-Ishi all contributed pictures and technical information for this book. The performance outlets of these manufacturers also helped. Don Hubbard authored a manual for Crane Cams detailing how to turbocharge an engine with Schwitzer Turbochargers. I borrowed many few installation tips and illustrations from this manual. Ak Miller and cohorts Jack Lufkin, Burke LeSage, Bill Edwards and Jon Meyer were more than cooperative.

George Spears of Shelby-Spearco and Spearco Performance provided many excellent pictures and helped me get pictures of his customers' installations. Bob Keller added much useful information. Gale Banks has been diligent in making superbly engineered marine installations. He also spent several days working on the cover photo.

Crown Manufacturing has been one of the pioneers in turbocharging engines and producing kits. Derek Torley has the same enthusiasm toward turbochargers as did founder Ted Trevor. Derek gave me numerous pictures and a lot of data.

John Gaspari sold Rajay Turbochargers all over the world for TRW International. He faithfully sent pictures and other information from many countries.

M & W Gear is an important manufacturer in marine turbocharging. Jack Bradford, their Chief Engineer, has been turbocharging diesel and gasoline engines since about 1960 and really knows how to make complete bolt-on kits.

Tom Scahill became interested in turbocharging when he helped Howard Arneson install them on his off-shore racing boat. Since then, Tom has established a wonderful reputation doing custom installations.

Doug Roe's hints about installations will be helpful to the old timer as well as the novice.

Jim Kinsler of Kinsler Fuel Injection provided much information and many photos on racing installations he has done in recent years. McLaren Engines also helped with information and photos.

Reeves Callaway of Callaway Turbosystems responded quickly to requests for information and photos of his high-quality kits and components.

Lou Cruse of AiResearch, and Andy Johnston and Jon Meyer of Roto-Master helped by giving the manuscript a final read before the book was declared ready for the printer.

Installation Drawings

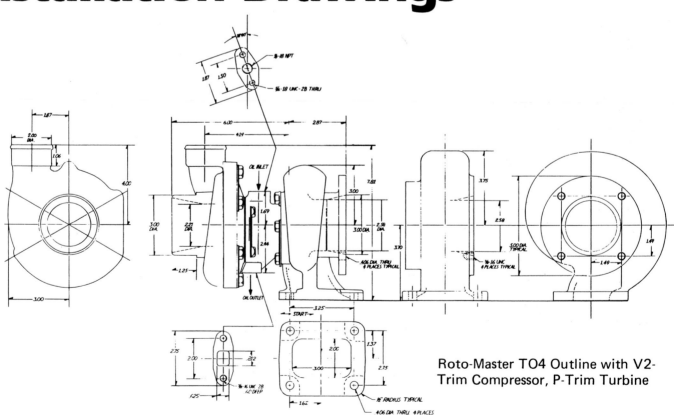

Roto-Master TO4 Outline with V2-Trim Compressor, P-Trim Turbine

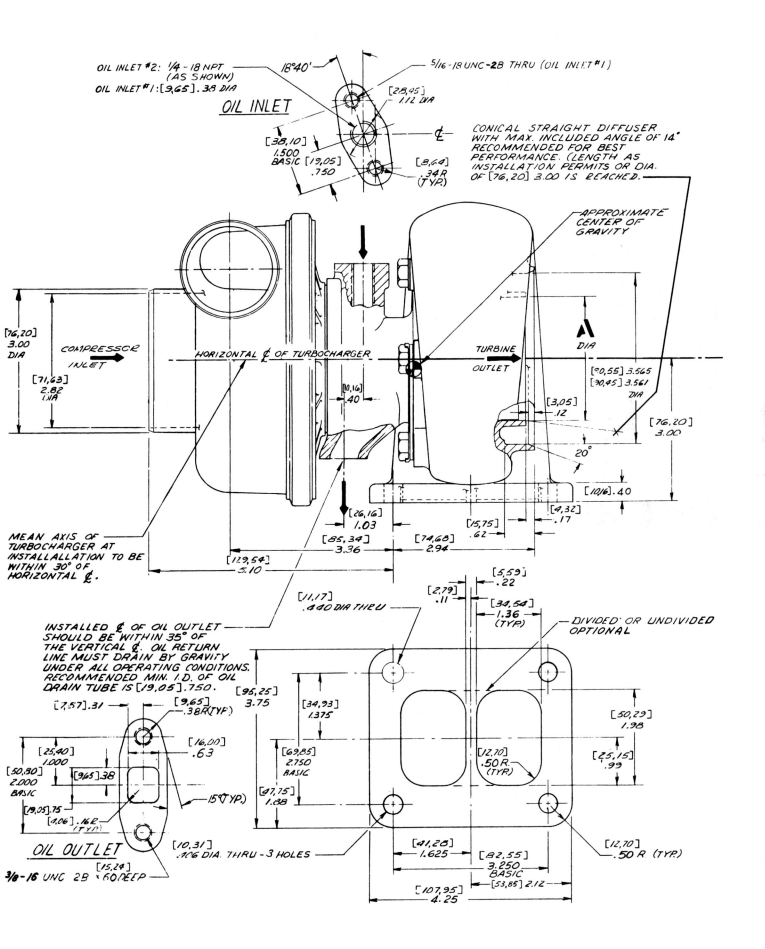

OIL INLET #2: 1/4-18 NPT
(AS SHOWN)
OIL INLET #1: [9,65] .38 DIA

18°40'

5/16-18 UNC-2B THRU (OIL INLET #1)

OIL INLET

[28,45] 1.12 DIA

[38,10] 1.500 BASIC [19,05] .750

[8,64] .34 R (TYP.)

CONICAL STRAIGHT DIFFUSER
WITH MAX. INCLUDED ANGLE OF 14°
RECOMMENDED FOR BEST
PERFORMANCE. (LENGTH AS
INSTALLATION PERMITS OR DIA.
OF [76,20] 3.00 IS REACHED.

APPROXIMATE
CENTER OF
GRAVITY

[76,20] 3.00 DIA

COMPRESSOR INLET

HORIZONTAL ₵ OF TURBOCHARGER

TURBINE OUTLET

A DIA

[71,63] 2.82 DIA

[90,55] 3.565
[90,45] 3.561
DIA

[3,05] .12

[76,20] 3.00

20°

[10,16] .40

[0,16] .40

[3,05] .12

MEAN AXIS OF
TURBOCHARGER AT
INSTALLALLATION TO BE
WITHIN 30° OF
HORIZONTAL ₵.

[26,16] 1.03

[15,75] .62

[4,32] .17

[85,34] 3.36

[74,68] 2.94

[10,16] .40

[129,54] 5.10

[5,59] .22

[2,79] .11

[34,54] 1.36 (TYP.)

INSTALLED ₵ OF OIL OUTLET
SHOULD BE WITHIN 35° OF
THE VERTICAL ₵. OIL RETURN
LINE MUST DRAIN BY GRAVITY
UNDER ALL OPERATING CONDITIONS.
RECOMMENDED MIN. I.D. OF OIL
DRAIN TUBE IS [19,05] .750.

[11,17] .440 DIA THRU

DIVIDED OR UNDIVIDED
OPTIONAL

[7,57] .31

[9,65] .38 R (TYP.)

[16,00] .63

[95,25] 3.75

[34,93] 1.375

[50,29] 1.98

[25,40] 1.000

[9,65] .38

[69,85] 2.750 BASIC

[12,70] .50 R. (TYP.)

[25,15] .99

[50,80] 2.000 BASIC

[19,05] .75

[47,75] 1.88

[4,06] .16 R (TYP.)

15° (TYP.)

OIL OUTLET

[10,31] .106 DIA. THRU - 3 HOLES

[41,28] 1.625

[82,55] 3.250 BASIC

[12,70] .50 R. (TYP.)

3/8-16 UNC 2B × .60 DEEP

[15,24]

[53,85] 2.12

[107,95] 4.25

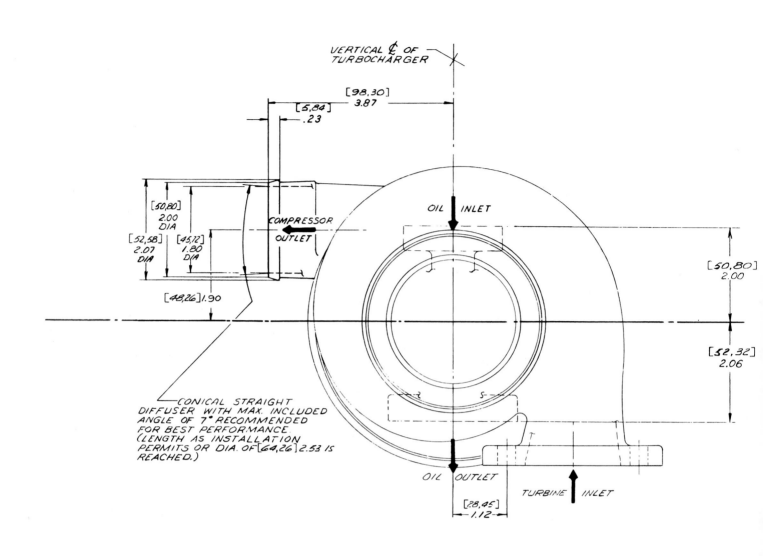

VERTICAL ℄ OF
TURBOCHARGER

[98,30]
3.87

[5,84]
.23

[50,80]
2.00
DIA

[52,58]
2.07
DIA

[45,72]
1.80
DIA

COMPRESSOR
OUTLET

[48,26] 1.90

OIL INLET

[50,80]
2.00

[52,32]
2.06

CONICAL STRAIGHT
DIFFUSER WITH MAX. INCLUDED
ANGLE OF 7° RECOMMENDED
FOR BEST PERFORMANCE.
(LENGTH AS INSTALLATION
PERMITS OR DIA. OF [64,26] 2.53 IS
REACHED.)

OIL OUTLET

TURBINE INLET

[28,45]
1.12

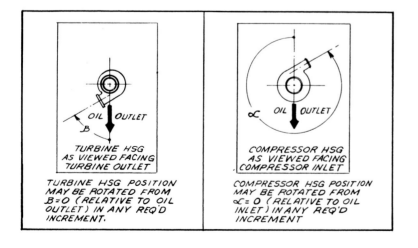

OIL OUTLET

B

TURBINE HSG
AS VIEWED FACING
TURBINE OUTLET

TURBINE HSG POSITION
MAY BE ROTATED FROM
B=0 (RELATIVE TO OIL
OUTLET) IN ANY REQ'D
INCREMENT.

α

OIL OUTLET

COMPRESSOR HSG
AS VIEWED FACING
COMPRESSOR INLET

COMPRESSOR HSG POSITION
MAY BE ROTATED FROM
α=0 (RELATIVE TO OIL
INLET) IN ANY REQ'D
INCREMENT.

**STANDARD TURBINE
HOUSING SIZES**

A/R

.68
.81
.96

AiResearch TO4 Turbocharger Outli

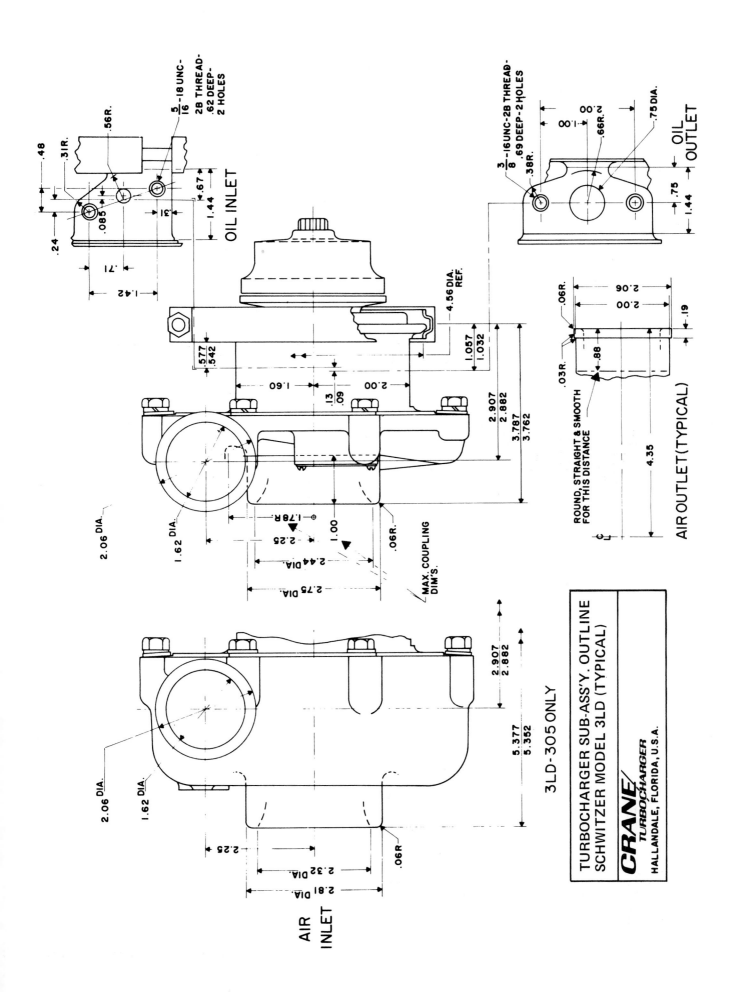

OIL INLET

$\frac{5}{16}$–18 UNC–
2B THREAD–
.62 DEEP–
2 HOLES

.56R.

.31R.

.48

.085

.24

.71

1.42

.31

.67

1.44

OIL OUTLET

$\frac{3}{8}$–16UNC–2B THREAD–
.69 DEEP–2 HOLES

.38R.

2.00

1.80

.66R.

.75 DIA.

.75

1.44

2.06

2.00

.06R.

.03R.

.88

.19

4.35

ROUND, STRAIGHT & SMOOTH
FOR THIS DISTANCE

AIR OUTLET (TYPICAL)

4.56 DIA.
REF.

1.057
1.032

2.907
2.882

3.787
3.762

.577
.542

1.60

.13

.60

2.00

2.06 DIA.

1.62 DIA.

2.25

1.78R.

1.00

2.44 DIA.

2.75 DIA.

.06R.

MAX. COUPLING
DIM'S.

2.907
2.882

5.377
5.352

3LD–305 ONLY

2.06 DIA.

1.62 DIA.

2.25

2.32 DIA.

2.81 DIA.

.06R.

AIR
INLET

TURBOCHARGER SUB-ASS'Y. OUTLINE
SCHWITZER MODEL 3LD (TYPICAL)

CRANE
TURBOCHARGER
HALLANDALE, FLORIDA, U.S.A.

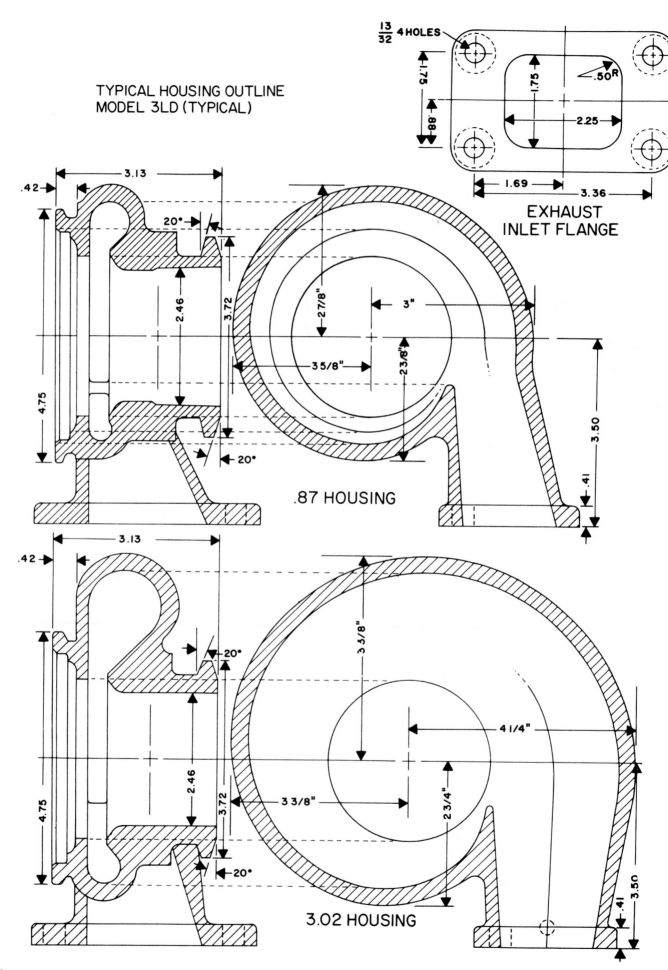

TYPICAL HOUSING OUTLINE
MODEL 3LD (TYPICAL)

EXHAUST
INLET FLANGE

.87 HOUSING

3.02 HOUSING

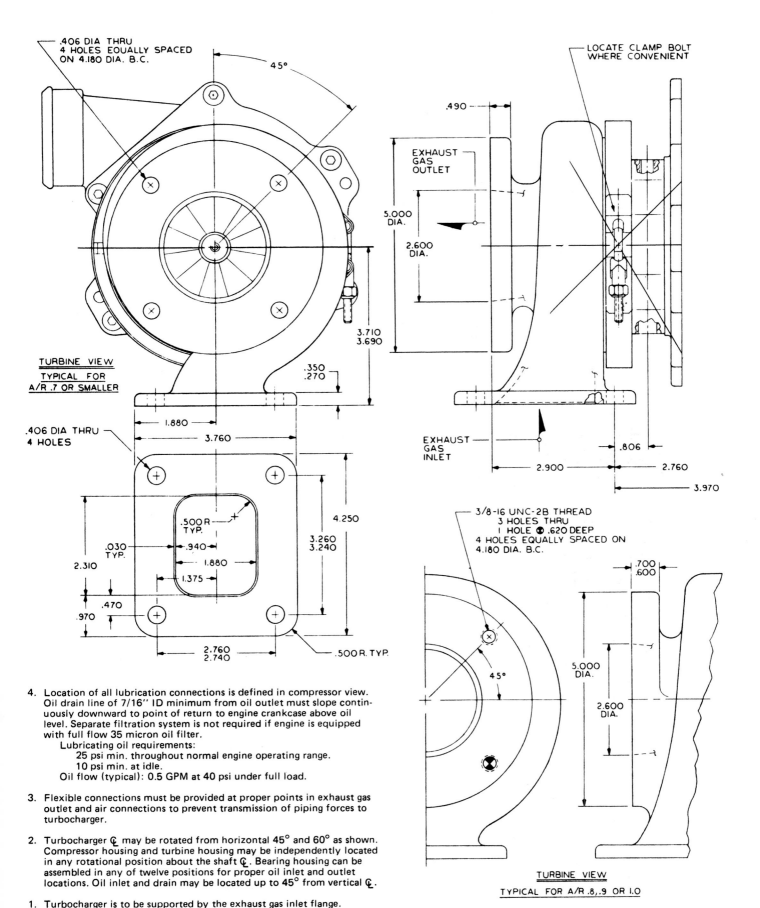

.406 DIA THRU
4 HOLES EQUALLY SPACED
ON 4.180 DIA. B.C.

45°

TURBINE VIEW
TYPICAL FOR
A/R .7 OR SMALLER

.406 DIA THRU
4 HOLES

3.710
3.690

.350
.270

1.880

3.760

.500 R
TYP.

.030
TYP.

.940

1.880

1.375

4.250

3.260
3.240

2.310

.470

.970

2.760
2.740

.500 R. TYP.

LOCATE CLAMP BOLT
WHERE CONVENIENT

.490

EXHAUST
GAS
OUTLET

5.000
DIA.

2.600
DIA.

EXHAUST
GAS
INLET

.806

2.900

2.760

3.970

3/8-16 UNC-2B THREAD
3 HOLES THRU
I HOLE ⊕ .620 DEEP
4 HOLES EQUALLY SPACED ON
4.180 DIA. B.C.

.700
.600

45°

5.000
DIA.

2.600
DIA.

TURBINE VIEW
TYPICAL FOR A/R .8,.9 OR 1.0

4. Location of all lubrication connections is defined in compressor view. Oil drain line of 7/16″ ID minimum from oil outlet must slope continuously downward to point of return to engine crankcase above oil level. Separate filtration system is not required if engine is equipped with full flow 35 micron oil filter.
    Lubricating oil requirements:
       25 psi min. throughout normal engine operating range.
       10 psi min. at idle.
    Oil flow (typical): 0.5 GPM at 40 psi under full load.

3. Flexible connections must be provided at proper points in exhaust gas outlet and air connections to prevent transmission of piping forces to turbocharger.

2. Turbocharger ℄ may be rotated from horizontal 45° and 60° as shown. Compressor housing and turbine housing may be independently located in any rotational position about the shaft ℄. Bearing housing can be assembled in any of twelve positions for proper oil inlet and outlet locations. Oil inlet and drain may be located up to 45° from vertical ℄.

1. Turbocharger is to be supported by the exhaust gas inlet flange.

151

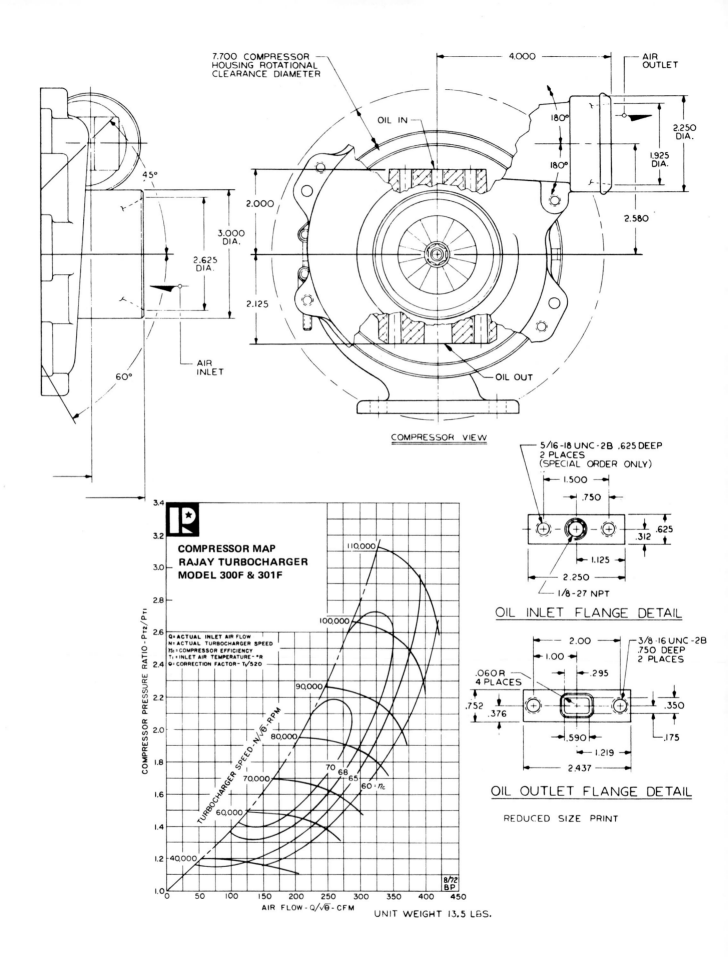

7.700 COMPRESSOR HOUSING ROTATIONAL CLEARANCE DIAMETER

AIR OUTLET

4.000

OIL IN

180°

180°

2.250 DIA.

1.925 DIA.

2.000

2.580

45°

3.000 DIA.

2.625 DIA.

2.125

60°

AIR INLET

OIL OUT

COMPRESSOR VIEW

5/16-18 UNC-2B .625 DEEP 2 PLACES (SPECIAL ORDER ONLY)

1.500

.750

.625

.312

1.125

2.250

1/8-27 NPT

OIL INLET FLANGE DETAIL

2.00

1.00

.295

3/8-16 UNC-2B .750 DEEP 2 PLACES

.060 R 4 PLACES

.752

.376

.350

.175

.590

1.219

2.437

OIL OUTLET FLANGE DETAIL

REDUCED SIZE PRINT

COMPRESSOR MAP RAJAY TURBOCHARGER MODEL 300F & 301F

Q = ACTUAL INLET AIR FLOW
N = ACTUAL TURBOCHARGER SPEED
η_c = COMPRESSOR EFFICIENCY
T_1 = INLET AIR TEMPERATURE - °R
θ = CORRECTION FACTOR - $T_1/520$

COMPRESSOR PRESSURE RATIO - P_{T2}/P_{T1}

TURBOCHARGER SPEED - $N/\sqrt{\theta}$ - RPM

110,000

100,000

90,000

80,000

70,000

60,000

40,000

70 68 65 60 : η_c

8/72 BP

AIR FLOW - $Q/\sqrt{\theta}$ - CFM

UNIT WEIGHT 13.5 LBS.

152

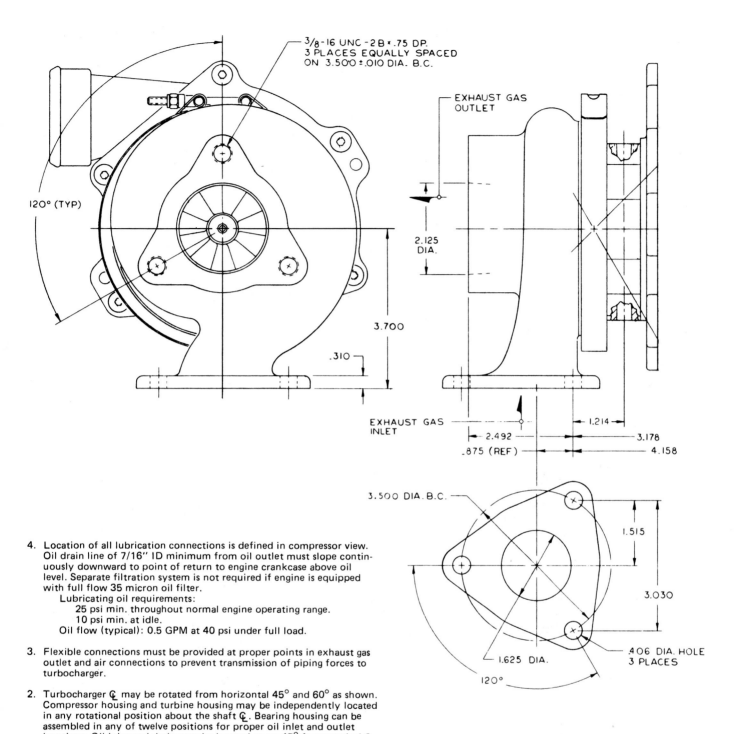

3/8-16 UNC-2B × .75 DP.
3 PLACES EQUALLY SPACED
ON 3.500 ±.010 DIA. B.C.

EXHAUST GAS OUTLET

120° (TYP)

2.125 DIA.

3.700

.310

EXHAUST GAS INLET

2.492

.875 (REF)

1.214

3.178

4.158

3.500 DIA. B.C.

1.515

3.030

.406 DIA. HOLE
3 PLACES

1.625 DIA.

120°

4. Location of all lubrication connections is defined in compressor view. Oil drain line of 7/16" ID minimum from oil outlet must slope continuously downward to point of return to engine crankcase above oil level. Separate filtration system is not required if engine is equipped with full flow 35 micron oil filter.
   Lubricating oil requirements:
   25 psi min. throughout normal engine operating range.
   10 psi min. at idle.
   Oil flow (typical): 0.5 GPM at 40 psi under full load.

3. Flexible connections must be provided at proper points in exhaust gas outlet and air connections to prevent transmission of piping forces to turbocharger.

2. Turbocharger ℄ may be rotated from horizontal 45° and 60° as shown. Compressor housing and turbine housing may be independently located in any rotational position about the shaft ℄. Bearing housing can be assembled in any of twelve positions for proper oil inlet and outlet locations. Oil inlet and drain may be located up to 45° from vertical ℄.

1. Turbocharger is to be supported by the exhaust gas inlet flange.

153

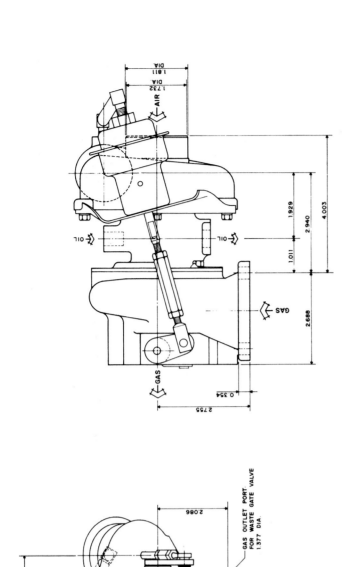

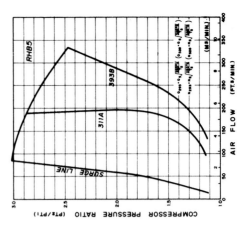

WARNER-ISHI
707 SOUTHSIDE DRIVE
DECATUR, ILLINOIS 62525
PHONE: (217) 428-4631

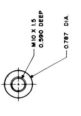

M10 X 1.5
0.590 DEEP
0.787 DIA.

OIL INLET FLANGE

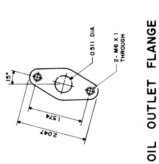

0.511 DIA.
2 - M6 X 1
THROUGH

OIL OUTLET FLANGE

GAS OUTLET PORT
FOR WASTE GATE VALVE
1.377 DIA.

GAS OUTLET PORT
FOR TURBINE
1.929 DIA.

4 - M8 X 1.25
0.511 DEEP

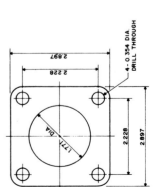

4 - 0.354 DIA
DRILL THROUGH

1.771 DIA.

GAS INLET FLANGE

154

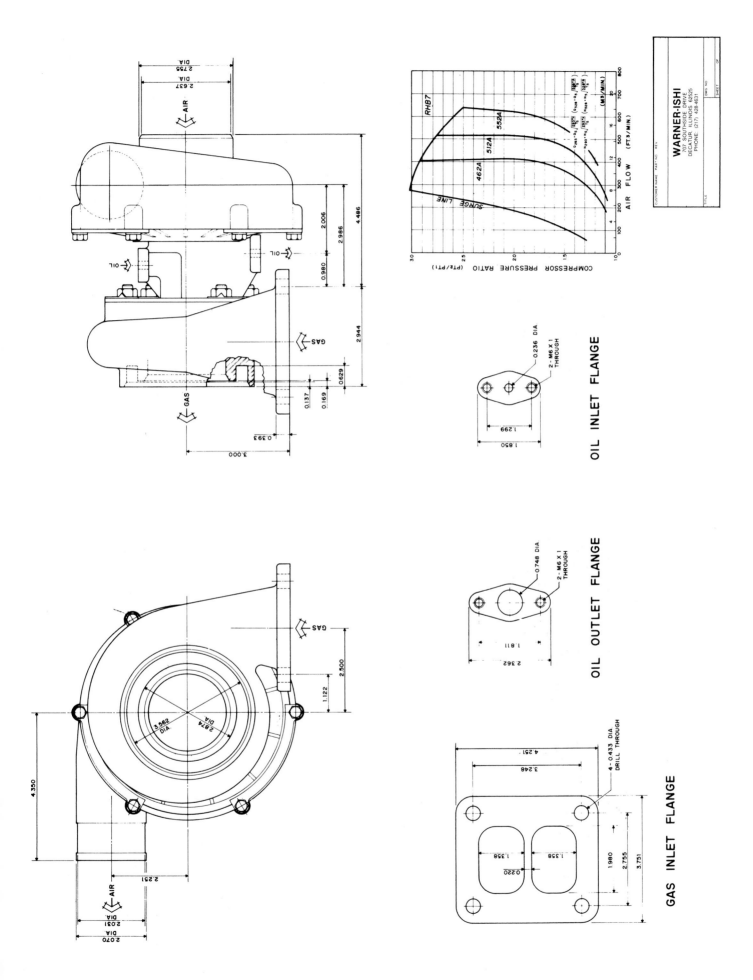

OIL INLET FLANGE

OIL OUTLET FLANGE

GAS INLET FLANGE

WARNER-ISHI
707 SOUTHSIDE DRIVE
DECATUR, ILLINOIS, 62525
PHONE: (217) 428-4631

155

Notes

Notes

Index

HANDBOOKS

Auto Electrical Handbook: 0-89586-238-7
Auto Upholstery & Interiors: 1-55788-265-7
Brake Handbook: 0-89586-232-8
Car Builder's Handbook: 1-55788-278-9
Street Rodder's Handbook: 0-89586-369-3
Turbo Hydra-matic 350 Handbook: 0-89586-051-1
Welder's Handbook: 1-55788-264-9

BODYWORK & PAINTING

Automotive Detailing: 1-55788-288-6
Automotive Paint Handbook: 1-55788-291-6
Fiberglass & Composite Materials: 1-55788-239-8
Metal Fabricator's Handbook: 0-89586-870-9
Paint & Body Handbook: 1-55788-082-4
Sheet Metal Handbook: 0-89586-757-5

INDUCTION

Holley 4150: 0-89586-047-3
Holley Carburetors, Manifolds & Fuel Injection: 1-55788-052-2
Rochester Carburetors: 0-89586-301-4
Turbochargers: 0-89586-135-6
Weber Carburetors: 0-89586-377-4

PERFORMANCE

Aerodynamics For Racing & Performance Cars: 1-55788-267-3
Baja Bugs & Buggies: 0-89586-186-0
Big-Block Chevy Performance: 1-55788-216-9
Big Block Mopar Performance: 1-55788-302-5
Bracket Racing: 1-55788-266-5
Brake Systems: 1-55788-281-9
Camaro Performance: 1-55788-057-3
Chassis Engineering: 1-55788-055-7
Chevrolet Power: 1-55788-087-5
Ford Windsor Small-Block Performance: 1-55788-323-8
Honda/Acura Performance: 1-55788-324-6
High Performance Hardware: 1-55788-304-1
How to Build Tri-Five Chevy Trucks ('55-'57): 1-55788-285-1
How to Hot Rod Big-Block Chevys:0-912656-04-2
How to Hot Rod Small-Block Chevys:0-912656-06-9
How to Hot Rod Small-Block Mopar Engines: 0-89586-479-7
How to Hot Rod VW Engines:0-912656-03-4
How to Make Your Car Handle:0-912656-46-8
John Lingenfelter: Modifying Small-Block Chevy: 1-55788-238-X
Mustang 5.0 Projects: 1-55788-275-4

Mustang Performance ('79–'93): 1-55788-193-6
Mustang Performance 2 ('79–'93): 1-55788-202-9
1001 High Performance Tech Tips: 1-55788-199-5
Performance Ignition Systems: 1-55788-306-8
Performance Wheels & Tires: 1-55788-286-X
Race Car Engineering & Mechanics: 1-55788-064-6
Small-Block Chevy Performance: 1-55788-253-3

ENGINE REBUILDING

Engine Builder's Handbook: 1-55788-245-2
Rebuild Air-Cooled VW Engines: 0-89586-225-5
Rebuild Big-Block Chevy Engines: 0-89586-175-5
Rebuild Big-Block Ford Engines: 0-89586-070-8
Rebuild Big-Block Mopar Engines: 1-55788-190-1
Rebuild Ford V-8 Engines: 0-89586-036-8
Rebuild Small-Block Chevy Engines: 1-55788-029-8
Rebuild Small-Block Ford Engines:0-912656-89-1
Rebuild Small-Block Mopar Engines: 0-89586-128-3

RESTORATION, MAINTENANCE, REPAIR

Camaro Owner's Handbook ('67–'81): 1-55788-301-7
Camaro Restoration Handbook ('67–'81): 0-89586-375-8
Classic Car Restorer's Handbook: 1-55788-194-4
Corvette Weekend Projects ('68–'82): 1-55788-218-5
Mustang Restoration Handbook('64 1/2–'70): 0-89586-402-9
Mustang Weekend Projects ('64–'67): 1-55788-230-4
Mustang Weekend Projects 2 ('68–'70): 1-55788-256-8
Tri-Five Chevy Owner's ('55–'57): 1-55788-285-1

GENERAL REFERENCE

Auto Math:1-55788-020-4
Fabulous Funny Cars: 1-55788-069-7
Guide to GM Muscle Cars: 1-55788-003-4
Stock Cars!: 1-55788-308-4

MARINE

Big-Block Chevy Marine Performance: 1-55788-297-5

HPBOOKS ARE AVAILABLE AT BOOK AND SPECIALTY RETAILERS OR TO
ORDER CALL: 1-800-788-6262, ext. 1

HPBooks
A division of Penguin Putnam Inc.
375 Hudson Street
New York, NY 10014